Male Great Crested Newt

Nigel Hand

Front cover:	Great crested newt	Will Watson
	Common toad	Simon Williams
	Adder	Nigel Hand
Back cover:	Common toad	Simon Williams

Herefordshire Amphibian and Reptile Team

Herefordshire Amphibian and Reptile Team (HART) was formed in February 2000 through an initiative of the Herefordshire Nature Trust and the Worcestershire Reptile and Amphibian Group. Since then HART has grown steadily and now operates independently though in close cooperation with its two 'parents'. HART also works in partnership with Herefordshire Farming and Wildlife Advisory Group (FWAG), Herefordshire Action for Mammals and the Herefordshire Biological Records Centre (HBRC).

The aims of HART are:

- to gather and collate amphibian and reptile records for the county;
- to promote public awareness of these species;
- to encourage conservation of their habitats;
- to train individuals in survey and identification skills.

HART's main activities include regular surveys and site monitoring, field trips in summer and indoor meetings in winter. HART organises winter work parties to undertake scrub and pond conservation management (also an enjoyable social activity) and training on habitat maintenance and creation.

To contact HART visit our website www.herefordhart.org
or contact the **Herefordshire Nature Trust 01432 356872**
who will direct you to the relevant people.

HART's Herefordshire Ponds and Newts Project is part-financed by the European Union's European Agricultural Guidance and Guarantee Fund and Defra through the Herefordshire Rivers LEADER+ Programme. Additional funds have also been received from the New Opportunities Fund SEED programme run by the Royal Society for Nature Conservation, the Environment Agency Wales and the English Nature Aggregates Levy Sustainability Fund through the Herefordshire Biodiversity Partnership.

Will Watson (signature)

Amphibians and Reptiles of Herefordshire

by
Nigel Hand
Will Watson
Phyl King

Photographs and illustrations
Nigel Hand
Phyl King
Will Watson
Simon Williams

Maps
Helen Forster, HBRC

Foreword

by Mark O'Shea

I spend a great deal of my time travelling to tropical countries in order to study their rich and diverse herpetofaunas but it has not always been that way. Throughout the 1970s and 1980s my idea of a great day out was to scour a local mineworkings in the Midlands for grass snakes or venture onto Kinver Edge to find adders, or Dorset for smooth snakes and sand lizards. I was not taking away, I was observing, counting, honing my photography skills, enjoying the encounter.

Circumstances have now taken me to over thirty countries, most of them tropical, and introduced me to hundreds of species of exotic reptiles and amphibians, but I have not forgotten my roots, the wonder of sighting an alert, coiled British snake in the early morning sun, or of finding great crested newts in terrestrial garb under debris around the edge of a forgotten village pond. These are childhood and adolescent memories that I treasure and ones that I would hope generations to come will be able to appreciate for themselves.

But they won't if the continued rate of habitat loss, fragmentation and alteration continues. The adder in particular is in trouble in the Midlands and adder-bashing may not be entirely to blame. This is a snake with very specific habitat and prey requirements. Change the surroundings and you may make its survival untenable. Neither does it deserve such treatment, one dozen fatal bites in the entire 20th century, come on, horses, dogs and bees kill many more people but we do not demonise them. We absolutely must protect our small but very special island herpetofauna.

This is why a publication like the one you hold in your hands, *Amphibians and Reptiles of Herefordshire*, is so important. National and regional field guides are all very good but conservation begins at the grass roots level and that means much more locally, by county or shire, even parish. And to conserve anything you first need to know what you have got and how many. Now, for Herefordshire, that information is available, painstakingly compiled by three dedicated authors with a passion for their home-grown snakes, lizards, frogs and newts.

This book concerns itself with five of Britain's native amphibians and four of its reptiles, but it also includes reports for three introduced species. Because the species numbers are not high the authors are able to devote a great deal more space to each species, providing much more information to enable the reader, not only to identify the frog, lizard or snake, but to understand it, what makes it tick, how it lives, to appreciate it as a wild animal. The excellent photographs included illustrate many aspects of the subjects' life histories.

Finally, for me one of the most important aspects of this book are the maps. I confess to being a bit of a cartophile (a map lover) so distribution maps are particular favourites since they combine two of my passions. The spot-marked distribution maps in this book are precise and detailed. Those for the amphibians look fairly healthy but unfortunately the reptile maps tell a different story and one that perfectly emphasises the urgent need for a local guide such as *Amphibians and Reptiles of Herefordshire*.

Contents

Introduction

by Richard King, Chairman of HART

Herefordshire has five of Britain's native amphibians: common frog and toad and smooth, palmate and great crested newt, and four of the native reptiles: viviparous (common) lizard, slow-worm, grass snake, and adder. They have suffered from loss of habitat over the last hundred years, and some have even been persecuted.

If we are to protect and conserve these species we need to have an accurate idea of where they are found in the county, and the state of their habitats. When formed in 2000 HART recognised that there were very few records of these species in the county and their distribution was poorly documented, so in 2003 it started the Herefordshire Ponds and Newts Project to survey ponds and pond life and map amphibian distribution. This book is the culmination of that process, showing the results of all the project surveys together with amphibian records from the past. It also includes the county's reptile records.

The book highlights the impact of the county's geology and landscape on the distribution of amphibians and reptiles and their habitats, and the way in which ponds, essential to amphibians, have evolved and been used by mankind over the centuries. There are sections on how recording has developed in the county, on conservation today, and on the importance of garden habitats. The species accounts give detailed information about life cycle and habitats, with descriptions to help identification. There are distribution maps for each species, with some reflections giving a local slant.

We are indebted to the authors for their energy and enthusiasm in their contribution to the conservation of these species. Will Watson, County Amphibian Recorder and Project Consultant, has set the standards for amphibian and pond recording, provided most of the training and a large number of the records. Nigel Hand, County Reptile Recorder, has produced most of the recent reptile records as he has almost single-handedly monitored the main reptile sites over the last few years. Phyl King has spent many hours revising and integrating the text and excellent photographs to put this book together.

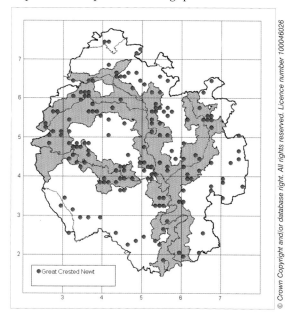

We hope the book will increase public understanding of these beautiful and fascinating creatures, and inspire people to do more to help conserve them and their habitats.

Herefordshire Ponds and Newts (HPN) Project 2003 - 2006

The aim of the project was to assess the health of the ponds in Herefordshire, and the distribution of the great crested newt and other

LEADER+ area showing concentration of great crested newt records

amphibians in the county. As the great crested newt is a Biodiversity Action Plan (BAP) priority species we had the opportunity to apply for funds for the work. The Herefordshire Rivers LEADER+ programme was just starting and agreed to fund a substantial percentage of the project. It was therefore restricted to the parts of Herefordshire for which LEADER+ funds could be used, namely parishes that bordered or included the four main Herefordshire rivers: the Wye, Lugg, Frome and Arrow. Will Watson and a team of volunteers have surveyed over 270 ponds in this area, generating an excellent record base for the county. The distribution maps for each of the newt species show a concentration of records in the LEADER+ area because of the concentration of effort there. This is highlighted on the map on page 6. The funding also allowed us to provide pond and amphibian training for over 150 people, thus expanding the availability of these skills in the county.

In the case of reptiles, records are fairly scarce and a similar project is needed to survey the county's reptiles on a more comprehensive basis. For all species, we need more records to improve still further our understanding of the status and distribution of amphibians and reptiles across the county. HART always welcomes more volunteer recorders, and records to add to the database.

Acknowledgements

We are very grateful to the LEADER+ Programme Team both for granting us the funds and for all the help and encouragement they have given us over the course of the project. We also thank the other organisations which granted us funds, and the Herefordshire Nature Trust for its full support, guidance and advice in the preparation of the project definition and funding application and during the project itself, including the use of its headquarters for meetings and general resource support.

We thank the pond owners who gave us access to their ponds, the survey team coordinators who organised training and support, and the volunteers who put in so much time and effort surveying ponds. We thank the environmental consultancies and the Herefordshire Council for ensuring that their records were released to the HBRC, and thanks also to all other recorders whose records are included in the maps. The publication of this book would not have been possible without the extensive help and guidance received from the Herefordshire Biological Records Centre and its manager Steve Roe. We are grateful to Helen Forster of the Centre for creating the species distribution maps, and the volunteers who entered the large amount of data collected.

We are particularly grateful to Mark O'Shea, international herpetologist and broadcaster, for kindly supporting us by writing the Foreword, to John Baker of the Herpetological Conservation Trust for his expert advice and assistance in checking the text, and to Francesca Griffith, Jo Hackman, Margaret Wrenn and members of the HBRC for the careful process of proof reading. We are grateful to Alastair Macdonald of the Royal (Dick) School of Veterinary Studies, University of Edingburgh for providing the photo of Gerald Leighton, and to the family of Dave Green, who sadly died in 2004, for giving us permission to print his drawings of the development of the great crested newt tadpole.

A list of all those who have contributed may be found at the end of the book.

We would like to give particular credit to HART's chairman, Richard King, for masterminding the HPN Project and the production of this book.

Geology, Landscape and Ponds of Herefordshire

Geology

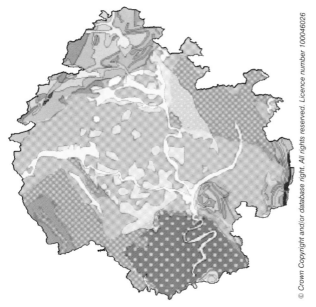

Superficial		
		alluvium
		glaciofluvial
Bedrock		
Lower Old Red Sandstone	Brownstones	
	St Maughans	
	Raglan Mudstone	
Silurian	Ludlow	
	Wenlock Limestone	
	Precambrian Granite, diorite, tonalite	

Approximate ages of the rocks mentioned in the text

Period	Millions of years ago
Quaternary	1.8 - present
Devonian	416 - 359
Silurian	444 - 416
Cambrian	542 – 488
Precambrian	Before 542

Geology Map of Herefordshire
IPR/79-25C British Geological Survey © NERC. All rights reserved

The bedrock underlying about four fifths of Herefordshire is Lower Old Red Sandstone (all the areas coloured orange on the map), consisting mainly of desert-derived sandstone and mudstone which gives rise to the county's familiar red soil. It comprises three geological formations, Raglan Mudstone, St Maughans and Brownstones.

The Raglan Mudstone Formation of the central plain dates from the Upper Silurian Period and consists of mudstones with smaller amounts of sandstone and limestone which on weathering break down into sandy silts, shales and clay. The clay tends to concentrate on lower ground and in the river valleys, and may be augmented by alluvial clays on the river flood plains. So lower lying parishes have a greater density of ponds because of both the predominant clay and the higher water table. These ponds are typically between 100 and 2500 m^2 in area.

The St Maughans Formation, dating from the Lower Devonian Period, is on the higher ground in the north-east and south of the county and contains sandstones, mudstones and calcretes, but has a greater concentration of the harder sandstones and exposed rocks. Larger groundwater-fed ponds, typically located in valleys, are more common and smaller catchment-fed ponds less so than on the Raglan Mudstone. For example on the freer draining ground of the Bromyard Plateau the majority of the ponds are between 2500 and 5000 m^2. Despite the presence of the sand most pond water is close to neutral. The water chemistry seems to suit palmate newts which are found in a wide variety of pond sites on these substrates whilst they remain absent or rare from neutral and mildly basic ponds on

Phyl King

Pool at Stretton Sugwas gravel pits

Will Watson

The Lawn Pool, an important natural pool at Moccas Park

Phyl King

View across the Wye Valley to the Woolhope Dome

other clay substrates in our region. Many of the semi-permanent ponds over sandstone support all three newt species. The Brownstones occupy the higher ground between Hereford and Ross and are also rich in sandstone but have little or no calcareous rock.

The oldest rocks of the county are the Pre-Cambrian and Cambrian rocks of the Malvern Hills, the former igneous in origin, hard and crystalline. Most ponds have formed as a result of quarrying. The water is mildly acidic and all five species of amphibian occur.

Older Silurian rocks are restricted to five areas, covering less than 5% of the county, of which the largest locations are the Woolhope Dome and the west flank of the Malvern Hills. Ponds are at low density here.

Carboniferous rocks are poorly represented and there are few ponds on this substrate.

Glacial deposits of the Quaternary Period are mostly of the late Devensian age and associated with the most recent glaciation about 26 to 14 thousand years ago, although in the lower Lugg valley there are the remains of four gravel terraces which pre-date the Devensian. Prior to the last glaciation the Lugg was joined by the Teme with the result that the valley of the Lugg is now wider than that of the Wye. In some places between Stretton Sugwas and Hereford extensive deposits of sand and gravel were laid down as the Wye valley ice retreated. Sand and gravels have been exploited commercially at Bodenham, Wellington, Withington and Stretton Sugwas. Man-made pools and lakes now exist as a consequence of these excavations.

As glaciers retreated large blocks of ice buried under glacial debris melted forming natural lakes and pools called kettle holes. Many of these still exist, most notably beside the River Wye between Hay and Hereford, although they are often modified by enlargement or 'restoration'. The Lawn Pool at Moccas Park is one of the best known of these natural landscape features.

Landscape

Herefordshire, with an area of 2156 km², just 1.69% of the total land area of the United Kingdom, borders the counties of Shropshire, Worcestershire, Gloucestershire, Powys and Gwent. Less than 2% of the county is urbanised, and it is recognised as being one of the most rural counties in England having almost everything we associate with the traditional English countryside: woodlands, orchards, clear running rivers, ancient field systems with tall hedges, valleys and vistas of rising hills, and a wealth of black and white timber framed buildings.

Central Herefordshire is predominantly lowland, punctuated by plateaus and groups of isolated flat-topped hills that have been shaped by four rivers: the Wye, Lugg, Arrow and Frome. The Bromyard Plateau (with Hegdon Hill the highest point at 254 m) covers nearly a quarter of the county and is incised by the Frome Valley and smaller river valleys. The Woolhope plateau to the east of Hereford is composed of hard layers of limestone alternating with softer shales, which have been pushed up by earth movements into a dome. The Woolhope Dome rises abruptly from the floodplain of the rivers Wye and Lugg and covers about 45 km^2. The most prominent flat-topped hills are the heavily-wooded Dinmore and Birley Hills between Hereford and Leominster, which rise steeply to 100 m above the Herefordshire Plain.

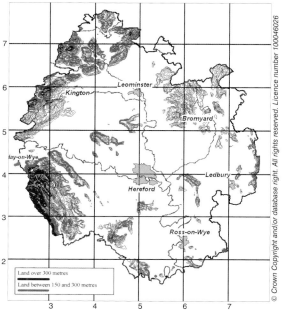

Contour map of Herefordshire

The county is bordered by high ground. The spectacular table-top plateau of the Black Mountains to the south west, at 700 m, is the highest point in the county. The region adjoining the Black Mountains, known as the Golden Valley, includes the prominent foothills Cusop Hill, Little Mountain, Merbach Hill, Garway Hill and Cefn Hill. To the south and east of the county are May Hill and the Forest of Dean, both just over the border in Gloucestershire. East are the Malvern Hills extending 12 km and rising to 397 m at the Worcestershire Beacon. To the north are the Mortimer Forest and the South Shropshire Hills, and north-west the foothills of the Radnorshire Forest, which include the Iron Age hill-forts at Wapley and Croft Ambrey. These hills are criss-crossed repeatedly by the River Teme, which cuts through a magnificent gorge at Downton.

The Malvern Hills and Mortimer Forest in particular, with their south and east facing slopes, wooded valleys and lowland ponds, provide good mixed reptile habitat, and all four of the county's reptile species occur. Some of the upland commons in the county are also good for reptiles, notably Yatton in the north and Ewyas Harold in the south-west.

Rivers are a key feature of the landscape. The Wye is the largest, rising in the Welsh mountains and flowing 130 miles through the

Horseshoe bend in the River Wye and floodplain ponds from Merbach Hill

county. The Lugg, Arrow and Teme also rise in Wales, while the Frome rises on the Bromyard plateau. The Teme flows into the Severn at Worcester, but the others converge and join the Wye east of Hereford, which is surrounded by the floodplains of the Wye and Lugg. The Wye then meanders south leaving the county just before Monmouth. Herefordshire's other main river, the Monnow, rises in the Black

The Malvern Hills

Mountains and for much of its course forms the border with Wales, eventually joining the Wye at Monmouth.

The fertile central Herefordshire Plain is intensively cultivated and is dominated by large arable fields, with fewer ponds, hedges and other wildlife habitats, although our commoner reptiles and amphibians still occur here. There is a marked contrast between this and the more traditional pastoral landscape typical of the west of the county, but even in the west the high density of grazing stock has led to declines in the undisturbed natural habitats which reptiles especially require.

In the urban environment slow-worms have been the most successful reptile to adapt, and frogs and newts have taken full advantage of the increase in garden ponds.

History of Herefordshire's Ponds

The most significant ponds within Herefordshire in terms of both geology and biology are kettle holes. They differ from other ponds in the county as they are natural features, mostly fed by groundwater. Their fluctuating water levels draw down considerably in summer, which has a profound impact on pond flora and fauna. These large, relatively shallow pools, the largest of which is greater than two ha, are suitable breeding sites for great crested newts, as they dry out on occasions so eliminating fish which may colonise during flooding episodes. Many are located in the north-west quarter of the county between the Wye Valley west of Hereford and the Lugg Valley.

A series of large pools beside the River Wye includes Bumper Pool, the Withy Pool, Broadmoor Pool and the pool at Western. These four have been listed as Regionally Important Geological Sites (RIGS) by Herefordshire and Worcestershire Earth Heritage Trust. Natural ponds and pools are particularly special as they frequently support rare or unusual freshwater life, and contain uninterrupted sequences of sediment with preserved pollen and wood deposits which provide evidence of past climatic conditions and vegetation. They are also nationally scarce; it has been estimated that only 2% of ponds are of natural origin.

Of the many pools along the Wye Valley, the majority of which are likely to have originated as kettle holes, only a few remain in an unaltered natural state. On the floodplain at Letton Lake there are fragments of a glacial relic landscape which contains numerous small ponds. Pond density in the unimproved marshland at the Sturts Site of Special Scientific Interest (SSSI), owned by the Herefordshire Nature Trust, reaches 20 per

Reservoir at Pipe and Lyde which has a good population of great crested newts

km². Parishes in the north-west of the county with large numbers of glacial relic ponds include Titley, Staunton-on-Arrow and Shobdon. Flinsham and Titley Pool in Titley parish have been designated as SSSIs for their geomorphological importance.

In Herefordshire, as in other English counties, ponds have been created for a variety of uses. The majority of farms within the county would have had several ponds, mainly constructed for ducks and/or fish. Where the ponds were located close to buildings they would also serve in emergencies as fire ponds; especially important when buildings were of wooden construction and there was no fire service. Some spring-fed ponds supplied drinking water into the early part of the twentieth century.

Most of the smaller ponds on the farm were dug for watering stock and horses. Cattle in particular require a regular and reliable water supply, and horses working the land needed a local supply of water for drinking. Ponds were often constructed in the corners of pastures, with arms leading out into two or more fields or abutting a hedge. Many of Herefordshire's farm ponds have stone lined bases which assisted stock and horse access and would have facilitated periodic de-silting.

Herefordshire has never had as many ponds as the neighbouring counties of Worcestershire and Gloucestershire. A study by Anthea Brian and Beryl Harding (1996) estimated that in the 1920s there were 5,231 ponds in the county compared with 7,653 in similar sized Worcestershire. Ponds could be unreliable sources of water, drying up in mid-summer when

One of the natural ponds at the Sturts SSSI

most needed, so more often than not either large ponds had to be constructed or modified for stock, or stock was given access to flowing water often with purpose-built drinking bays. The same study also noted the relative abundance of large ponds in the county, and found that 50% of ponds were formed by the damming of streams.

The Sites & Monuments Record (Herefordshire Archaeology 2006) lists over 200 fish pools, which is by far the largest use for ponds in the register and suggests that Herefordshire people wanted their ponds to be productive. At Croft Castle, owned by the National Trust, a series of fish pools was constructed as a trout fishery in what is now known as Fishpool Valley SSSI. These larger pools, whilst of great benefit to toads and frogs, probably provided poor habitat for other amphibians, particularly the great crested newt, due to predation by fish.

Mineral extraction always plays an important role in clay landscapes and the Sites and Monuments Register has 93 references to the digging of clay. Many ponds were formed after the digging of pits for clay, which was used for daub, cooking pots, tiles and latterly for brick making. There were several major brickworks, most notably the Hampton Park Brick Works at Tupsley, which closed in 1937 but was responsible for creating the county's best site for the common frog. The smaller clay pit ponds which periodically dry up provide good habitat for newts.

The county has its fair share of moats with 127 listed in the Register and is one the most densely moated counties in the country. They were usually constructed for ornament rather than defence, although the moat at Bronsil Castle is clearly a defensive feature. Lower Brockhampton, near Bromyard, is one of the finest examples of a moated

Drovers' pond at Fownhope which is a breeding site for all five of the county's amphibians

Will Watson

medieval manor house in England. Medieval moats, such as the one at the Court of Noke, were transformed into water features in the 17th century. The great period of the country house was in the 18th and 19th centuries, and ornamental lakes and pools were constructed on many of the larger estates. At Berrington Hall (National Trust), a pool with an island fed by a tributary of the River Lugg was created by Capability Brown, and at Eastnor Castle in the 19th century a lake was constructed as a backdrop to the castle.

There were numerous drovers' ponds sited along highways to provide a regular water supply to stock going to market. Several can be seen along the A44 between Docklow and Bredonbury. There are relatively few village ponds because the county has hardly any village greens. Examples that do exist include the ponds on Garway Common, the pond at Ashperton, which is on a tiny piece of public land, and the Lough Pool at Sellack.

Herefordshire, along with other lowland English counties, has lost large numbers of ponds. Some ponds and wetlands suitable for amphibian breeding were lost as a consequence of early drainage schemes, the earliest occurring in Roman times. With the onset of mechanisation in the 18th and 19th centuries the rate of change accelerated, and after the mid 20th century pond numbers suffered a major decline. There was an estimated loss of 30% of ponds within the county from the 1920s to the 1980s (Brian and Harding 1996), mostly in the fertile arable farmland of the central plain, with mainly smaller field ponds being affected. However greater interest in ponds and conservation since the 1990s is an encouraging development.

Conservation: Policy and Practice

In the second half of the 20th century there were substantial losses of key habitats for our amphibians and reptiles: not only ponds, but also unimproved grassland, hedgerows and other wildlife corridors, rough and untidy areas on farms, and scrub covered hillside. Many of these losses came from changes in farming practice as a result of government and European policy, and from the introduction of new technologies. For instance, many field ponds were filled in when farmers were encouraged to replace ponds with cattle troughs supplied by piped water to prevent the possible spread of water-borne diseases, or were destroyed because modern machinery allowed more pasture to be converted to arable. Water quality also declined in this period contributing to the impoverishment of wetland habitats, as has simple neglect leading to ponds becoming shaded out by trees, and silted up with fallen leaves and vegetation. Whilst the countryside has suffered many losses, new habitats have been created through the recent enthusiasm for garden ponds.

But in recent years much greater awareness of the issues of loss of habitat and biodiversity has led to local action for conservation, and national and international pressure on governments to change agricultural policies in favour of wildlife. Countryside Stewardship, now replaced by the Entry and Higher Level Stewardship Schemes for farmers and landowners, has helped to halt and reverse the declines. Pond creation, including small reservoirs for crop irrigation, fishing pools, and pools for wildlife and private amenity use, together with buffer strips for wildlife around field margins and restoration of woodland, are beginning to have an effect. Herefordshire FWAG has been at the forefront in helping and encouraging farmers to take advantage of these schemes.

At the national level conservation organisations such as Pond Conservation, Froglife and the Herpetological Conservation Trust, have also played a major role in promoting the value of amphibian and reptile habitats, thereby helping reduce losses and encourage better management.

Herefordshire's Local Biodiversity Action Plans (LBAPs) have been developed as part of the UK Biodiversity Action Plan to stimulate and monitor active conservation of our threatened wildlife. The great crested newt and adder both figure in the Herefordshire plans, and the HPN Project has played a significant part in the implementation of the plan for the great crested newt by raising awareness of the value of ponds and giving advice to pond owners on restoration.

Action for Amphibians and Reptiles

Firstly, dig more ponds! Amphibians need water in which to breed but otherwise spend most of their lives on land, so good aquatic and terrestrial habitats are both vital for healthy populations to develop.

Farmers can restore ponds and create other habitat suitable for amphibians and reptiles including wildlife corridors by taking greater advantage of the environmental schemes on offer. Specialist advice can be sought from Herefordshire FWAG and other professionals in the field.

Those wanting to take a more active part can carry out surveys and habitat management with their local amphibian and reptile group. We still have much to learn about amphibian and reptile distribution in the county, and HART would welcome more records and more volunteers to undertake regular surveys and site monitoring. The Herefordshire Nature Trust also welcomes volunteers, as do other voluntary wildlife groups in the county.

Ponds are often at risk from unsympathetic developments. The Council's Planning Ecologist monitors all planning applications, but vigilance by members of the public to keep the Planning Authorities aware of any potential destruction of this sort is another way to help.

Encouraging Amphibians and Reptiles in the Garden

Creating a good wildlife pond in the garden is one of the best ways to help amphibians. Frogs and newts quickly discover and colonise new ponds. Amphibians generally occupy the vegetated margins of a pond, with only toads requiring a depth greater than a metre, so a pond should have several depths of water, with shallow sloping sides, an irregular outline and a deeper area in the middle to prevent it drying out.

A good variety of native plants in and around the pond is important in order to provide egg-laying sites, habitat for invertebrate prey, and cover both for protection and for ambushing prey. Suitable plants for the shallow areas include water forget-me-not *Myosotis scorpioides* and water mint *Mentha aquatica*, which are favourite egg laying plants for great crested newts, and in the deeper areas oxygenators such as hornwort *Ceratophyllum demersum* or spiked water-milfoil *Myriphyllum spicatum*. Invasive alien species such as New Zealand pygmyweed *Crassula helmsii* and water fern *Azolla filiculoides* must be avoided.

Heavy shading of a pond restricts the growth of aquatic plants on which amphibians depend, but some shading (less than 50% of the surface area) can be beneficial as it reduces evaporation, maintains a more even temperature and controls excessive vegetation so ensuring an area of open water. Accumulation of leaf litter reduces water quality which can be harmful to aquatic life, so leaves should be cleared out in winter.

On land amphibians forage for worms, slugs, snails and other invertebrates. So a zone around the pond of damp long grass, and piles of wood or rocks provide both protection and a hunting ground for prey as well as hibernacula. Wildlife corridors are needed to allow movement between ponds and their terrestrial habitats. Great crested newts can travel up to 250 m from a pond in search of food. Frogs and toads go much further.

Fish should be avoided, and ducks and herons discouraged, as they prey on the various life-stages of amphibians.

When planning to create or restore a pond consider consulting HART or other specialists to discover the significance of what is already

Phyl King

A new pond after just five years

17

there and for advice on amphibian and general wildlife requirements.

Detailed information on pond creation and restoration is obtainable from HART and HART's website, www.herefordhart.org, or from the publications listed at the end of this book.

The reptiles most likely to take up residence in gardens are slow-worms and grass snakes. Reptiles tend to be slow colonisers, so are only likely to be found in gardens next to woodland, churchyards, railway embankments, brownfield sites and allotments where populations already exist. Do not attempt to move reptiles to your garden.

A reptile friendly garden is ideally large, with areas of unmown or very lightly mown grass, plenty of sun, banks or hedges, rockeries and lots of vegetation, together with patches of brambles, nettles and shrubs. A succession of compost heaps can be attractive to both grass snakes and slow-worms. The bigger and warmer the heap the larger the numbers attracted. An amphibian pond may also attract grass snakes. A dry stone wall with many nooks and crannies and a good growth of vegetation nearby is an excellent reptile feature. Piles of logs or brash can create safe basking areas, and objects to hide under, such as sheets of old corrugated iron or pieces of carpet, can provide refuges under which reptiles can warm themselves in safety from predators.

No chemicals or pesticides should be used. Slow-worms themselves are effective pest controllers as they feed on slugs. Cats, dogs and free range chickens are detrimental to reptiles.

Gardeners are generally too tidy and need to learn to live with a few undisturbed areas. By following these simple guidelines you may be pleasantly surprised at what turns up.

Good garden habitat for amphibians and reptiles

Amphibian and Reptile Recording in Herefordshire

Although natural history was a popular pastime among Victorians, this interest did not seem to extend to the county's amphibians. There is only one amphibian record we are aware of from this pond: a Mr. Holdsworth of the Royal Zoological Society found palmate newts in abundance near the village of Letton in 1863 (Holdsworth, 1863). Specimens were sent by him to a meeting of the Zoological Society for exhibition.

The Woolhope Naturalists' Field Club is usually a reliable source of records but so far no records of amphibians have been found in the Transactions of the Club for the 19th and 20th centuries. A few records of amphibians have been made incidentally to other surveys, such as those in Anthea Brian and Beryl Harding's survey of Herefordshire Ponds (1996).

The first recent amphibian record is from the Herefordshire Nature Trust from 1977, for common frog in the Olchon Dingle. The 1980s brought the first significant numbers of records. In 1983 the National Amphibian Survey was set up by the then Nature Conservancy Council and Leicester Polytechnic which provided 24 records between 1983 and 1987. Anthea Brian and Beryl Harding's survey contributed 51 records between 1987 and 1991. In the 1990s there was an increase in the number of amphibian records being returned. Major contributors during this period were Ben Proctor who was the reptile and amphibian contact for the county, and the late Don Goddard, HART's first Chairman, who provided amphibian records whilst surveying pond invertebrates mostly on behalf of the National Trust.

HART was formed in 2000, and since then there has been more extensive and systematic recording under the guidance of the County Recorders Will Watson and Nigel Hand. The first survey for amphibians was carried out in 2000 in Hereford City by Jane Sweetman of the Herefordshire Nature Trust and HART.

Since 2000 there has been a large rise in the number of surveys undertaken by environmental consultants within the county; about 5% of all amphibian and reptile records originate from such sources.

In 2003 Ross Andrew, the Community Biodiversity Officer at the Herefordshire Nature Trust, engaged the public in recording plants and animals which were easily identifiable. Between 2003 and 2005 this generated over 350 records of common frog and common toad, which has made a major contribution to our understanding of the distribution of these species. Records came from all over the county, so the maps for frog and toad distribution show a fairly even spread, unlike the newts maps which show a concentration in the LEADER+ area (see page 7). Between 2004 and 2006 the HPN Project, which trained over a hundred volunteers, provided more than 550 amphibian records.

Between 2004 and 2006 the HPN Project, which trained over a hundred volunteers, provided more than 550 amphibian records.

The most significant sources of records and data for reptiles in Herefordshire are the Transactions of the Woolhope Naturalists' Field Club and Gerald Leighton's two books: *The Life History of British Serpents and their Local Distribution in the British Isles* (1901), which includes his MD Edinburgh thesis *The Reptilia of the Monnow Valley*, and *The Life History of British Lizards and their Local Distribution in the British Isles* (1903). These books provide interesting records and were ambitious for their time in collating the knowledge of all the respected naturalists of the period. Gerald Leighton lived at Grosmont, Pontrilas. By talking to gamekeepers and woodsmen he recorded sightings in his local area.

HART has no recent records for adders on Garway Hill, Leighton's main area of research, but the grass snake is now a frequent visitor to gardens in the area. This is the reverse of the situation a hundred years ago and may possibly be due to Leighton's overzealous collecting of specimens and encouragement of the local populace to bring him adders. To record a snake it was captured, killed and preserved in a bottle of formalin or alcohol. Thankfully this practice is now illegal and adders are captured only by digital photography!

Since that time, with the exception of the 1980s when Martin Noble contributed a large number of records for the Mortimer Forest area, records have been very sparse. The bulk of them, mainly from Nigel Hand, has been received since the formation of HART and the establishment of the HBRC in 2001.

Gerald Leighton MD

Grosmont (left) and Garway Hill (right) where Leighton carried out his studies of reptiles

Phyl King

Great Crested Newt *Triturus cristatus*

Male great crested newt in breeding condition

Description

The great crested newt is the largest of the three British species of newt, on average up to twice the size of the smooth and the slightly smaller palmate newts. It reaches a maximum adult length of 17 cm.

The upper body is a dark brownish black and is covered with warts, those along the sides of the body having white tips giving a speckled effect. The belly is bright yellowish

Terrestrial female (left) and male (right)

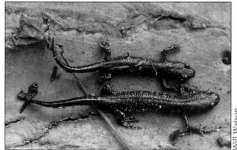

First and second year juveniles

orange with a striking pattern of black blotches unique to each individual. The toes are black with yellow or orange stripes.

In the breeding season the male develops a jagged crest along its back, which breaks at the rear of the abdomen before continuing to the end of the tail, and a very obvious wide silvery white stripe along the side of the tail. The female is usually larger than the male and in spring is noticeably swollen with eggs. The female has no crest and instead of the white tail stripe has a yellow one along the bottom of the tail which is present all year.

Life cycle

In common with the smaller newt species great crested newts typically spend late summer, autumn and winter on land, hiding in small mammal burrows, crevices in tree roots, or under rock or wood piles. In gardens they seek refuges under patios, in rockeries and old walls. Reports from people finding colonies of black looking lizards in gardens usually turn out to be great crested newts in hibernation. On damp mild evenings they may emerge from their resting places to forage on land. They prey on a range of invertebrates such as small earthworms, insects, spiders and slugs. They will also feed within their resting places.

They return to their breeding ponds on mild nights usually in March or April, occasionally as early as February. They may over-winter in ponds as adults, juveniles or tadpoles, but in our experience are the least likely of the three species to do so. During

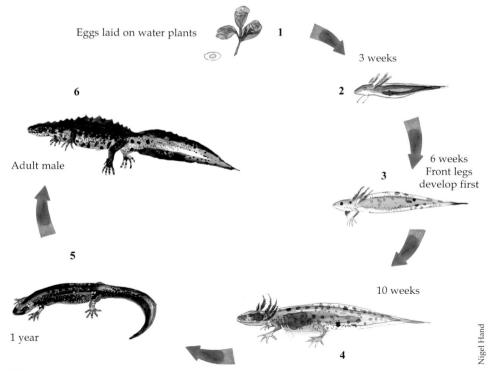

Eggs laid on water plants   **1**

3 weeks

6

2

6 weeks
Front legs
develop first

3

Adult male

5

10 weeks

1 year

4

Nigel Hand

Life cycle of the great crested newt

Breeding female

the day they hide in thick vegetation or in silt at the bottom of the pond, emerging into open water at night to feed and to perform their courtship displays. This is the best time to watch them with the aid of a torch. Adult newts whilst in the water feed on a wide variety of prey. In early spring they will take advantage of frog tadpoles, and in late April and May will feed on newt tadpoles and frequently cannibalise their own species.

Eggs are laid usually from March to May, the numbers laid each night building up as the year progresses, and a mature female may lay up to 200 in total by the end of the season. The eggs are laid on the leaves of water plants, which the female then folds over forming a protective pocket. Where several eggs are laid on one leaf this gives rise to an easily recognisable concertina effect. In the West Midlands the most frequently selected plants for egg laying are floating sweet-grass *Glyceria fluitans*, water forget-me-not *Myosotis scorpioides* and water mint *Mentha aquatica*. The whitish or cream coloured eggs are about 4.5 mm long, about double the size of the buff coloured eggs of the two smaller newt species.

Eggs laid on a single leaf of floating sweet-grass

Eggs on lesser water parsnip

Newly hatched tadpole

Great crested newt tadpole showing long slender toes and spotted tail fin

The feathery gills and broad head of the growing tadpole

Half of all great crested newt eggs die as the tail starts to develop due to a genetic anomaly. The surviving embryos hatch after about three weeks. The tadpoles have forward pointing gills, long slender toes, and a silvery underbelly. As they grow they develop a spotted tail fin which tapers to a fine point, and black spots on their flanks and upper surface which gradually coalesce into the continuous black markings typical of the adult. The head of the tadpole is the broadest part of the body. They are sometimes mistaken for small fish.

In their early stages the tadpoles feed mainly on water fleas and other micro-crustaceans, but as they grow they prey on a wide range of aquatic invertebrates including smaller newt tadpoles. After about 12 weeks the gills are gradually absorbed and the tadpoles start to gulp air from the surface of the pond. Around three weeks later they metamorphose into small

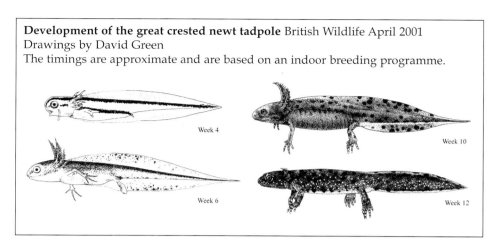

Development of the great crested newt tadpole British Wildlife April 2001
Drawings by David Green
The timings are approximate and are based on an indoor breeding programme.

Week 4

Week 10

Week 6

Week 12

versions of the adult, known as efts, and leave the water, usually between July and September. Males will return to water to breed after two or three years, females after three or four. They can move considerable distances over land, usually remaining within 500 m of the pond, although they can travel up to 1000 m. They do not necessarily return to the pond where they hatched, so readily colonise new ones in the vicinity. They are relatively long-lived, averaging seven to eight years.

Habitat

Great crested newts are generally found in larger, well established, nutrient rich ponds with open areas for display and plenty of weed for shelter, but they also occur in smaller garden ponds that are in the vicinity of established breeding populations. They avoid ponds with fish, which are significant predators of newt larvae, detecting their presence by smell. Ponds which draw-down rapidly in summer, or even dry up occasionally, can be favourable to great crested newts as desiccation eliminates fish.

Landscapes in which there are large numbers of ponds in close proximity can support robust great crested newt populations as the newts can move between ponds, colonising suitable ones as others become over-vegetated or dry out prematurely.

The importance of terrestrial habitat is often under-estimated. Habitats surrounding a pond needs to provide foraging and hibernation sites as well as dispersal corridors to other ponds, so features such as scrub, long grass, hedgerows, and piles of logs or stones in the immediate vicinity are important. Great crested newts hibernate under paving slabs, in old walls and in soil, often utilizing small mammal burrows. A study at one site revealed that the average hibernation depth was just 7 cm. Large populations will generally only establish where both the aquatic and terrestrial habitats meet their requirements.

Status, Threats and Legal Protection (2006)

The great crested newt is nationally the least common of our three species of newt and is the least able to adapt to changes in the landscape. Numbers declined significantly throughout most of the 20th century largely due to the infilling of ponds and the deterioration of both aquatic and surrounding habitats. Although pond losses have

Great Crested Newt
Triturus cristatus

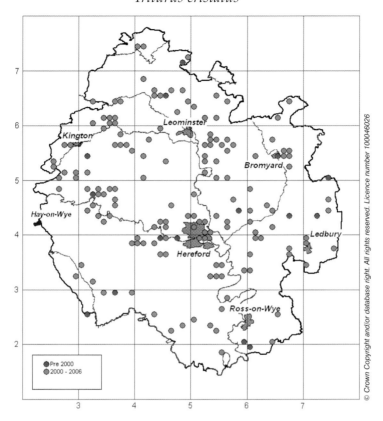

largely halted and new ponds are being constructed, many of these are managed for fish, and since fish prey on newt larvae these new ponds are of little or no value for breeding newts.

Because of its steep decline the great crested newt is protected under the European Habitats Directive 1992 and is also fully protected under the Wildlife and Countryside Act 1981 such that it is illegal to intentionally capture, kill, or injure them, or disturb them in their place of shelter. Great crested newt habitat is also protected and it is an offence to damage or destroy a breeding pond or resting place. A licence is required from Natural England or the appropriate statutory nature conservation agency to carry out surveys for scientific or educational purposes involving disturbance and handling.

The great crested newt has been selected as one of the 382 UK Biodiversity Action Plan species and has its own UK Species Action Plan. At a county level it is recognised as one of the 156 Priority species in the Local Biodiversity Action Plan (LBAP).

National and Local Distribution

Internationally lowland England and Wales are the main centres of great crested newt distribution; it is considered to be far rarer in mainland Europe. However its distribution across the United Kingdom is somewhat patchy and it is absent from Ireland and Cornwall and rare within Scotland. It is locally common in parts of North Wales, the West Midlands, Cheshire, and north west and south east England.

Great crested newts are widespread and locally common in Herefordshire except in the intensively managed agricultural land on the Herefordshire Plain where there are few ponds. The HPN Project has found that about 37% of the ponds surveyed support great crested newts. This is nationally a very high rate of pond occupancy and ranks Herefordshire alongside counties such as Worcestershire, Warwickshire, Cheshire, Gloucestershire, Kent, Sussex and Essex where they are also considered to be locally common.

In general great crested newts tend to thrive best in fairly open semi-permanent ponds with good vegetation structure, and usually avoid very large pools because of fish. Typically great crested newts are found in ponds with smooth newts, but in Herefordshire we have found that at least 30% of these ponds support palmate newts as well. This is 11% of all ponds surveyed by the Project; yet nationally occupation of a pond by all three species is a relatively unusual occurrence. Local conditions, particularly the geology of the county, have a bearing on this distribution. The coincidence map on page 28 shows the distribution of ponds with one or more species of newt and those ponds with all three species.

Unusual sites include ephemeral and permanent ponds on the Wye Valley floodplain between Hereford and Hay where we have recorded great crested newt tadpoles coexisting with mature three-spined stickleback *Gasterosteus aculeatus*. This is an unusual observation which requires further monitoring to understand the interrelationships.

We have so far identified few sites with good or exceptional populations. The only two sites where over a hundred adults have been recorded are a new farm reservoir to the north of Hereford city, about 1.2 ha in area and over 3 m deep, and three ponds located in a large wildlife garden in Llangarron.

Jagged crest with break at end of abdomen

Warty skin

Bright silvery-white tail stripe

Yellow and black striped toes

Vividly patterned yellow/orange and black belly

Nigel Hand

Identifying features of the male great crested newt

Coincidence map of great crested, smooth and palmate newts in ponds surveyed by the HPN Project

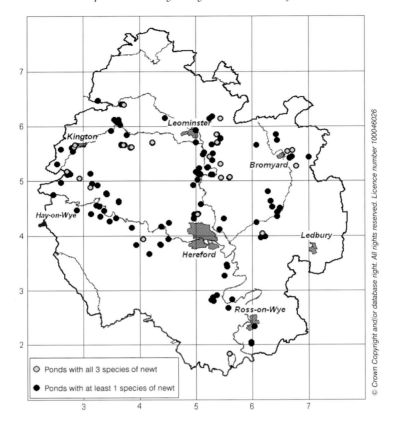

○ Ponds with all 3 species of newt

● Ponds with at least 1 species of newt

Smooth (Common) Newt *Triturus vulgaris*

Male smooth newt in breeding condition

Description

The smooth newt is the most common of Herefordshire's three newt species. At up to 10 cm long it is considerably smaller than the great crested newt. It is on average slightly larger than the palmate. Smooth and palmate newts frequently occur in the same ponds in the county and can be difficult to tell apart.

Both male and female smooth newt have a whitish underbelly which is speckled all over with dark spots. Both also have a whitish throat, which is usually spotted. In contrast the throat of the palmate newt is not spotted.

However, there are some marked differences between the sexes. The male's upper surface and flanks are a pale slatey-grey or light brown with dark brown blotches, and its underbelly has a central broad band of orange or yellowish orange covered in dark brown spots. In the breeding season the male develops a deep wavy crest along the back which extends without a break down the tail. There are small bands of orange and blue on the lower tail, nearest to the body. In peak breeding condition the male also develops flaps or fringes on the toes of the back feet. Both the orange underbelly and the crest in the male smooth newt can lead to confusion with the great crested newt, but the latter species has a serrated crest and is so much larger and bulkier that once seen there should be no further confusion.

The female is drabber than the male. The upper surface and flanks are typically a pale

Phyl King

Female smooth newt in breeding condition

Will Watson

Smooth newt, terrestrial phase

straw-colour which may have a brown tinge, and the tail has an orange line along its lower edge. It often looks smaller than the male as it has no crest.

In the terrestrial state the male is greyish-brown, with its orange belly pattern still visible, and the female is buff-brown. In hibernation the skin becomes velvety. Immatures are similar to females, and are difficult to distinguish from palmate immatures.

The differences between the breeding males of the smooth and palmate newts are highlighted in the drawings on page 37 of the palmate newt section.

Life cycle

Smooth newts over-winter in hibernacula on land and usually start moving to their breeding pools from February to March. This migration takes place over several weeks, so in general there are not the mass movements associated with frogs and toads, although on mild humid nights following a long cold spell large numbers may be on the move at the same time. Most return to land by June or July.

Male smooth newt. Note the fringes on the toes of the back feet. Both male and female have the spotted throat.

Phyl King

Males may initially outnumber females at the pond, so there can be competition for mates, but they are not territorial or aggressive. The male performs an elaborate courtship display, finally depositing a packet of sperm on the pond floor. He then guides the female over this spermatophore so that she can pick it up with her cloaca. Fertilisation of eggs is internal. The newts are more active in the pond at night especially around dawn and dusk, which is the best time to watch these courtship displays.

The female, over the season, can lay 100 to 600 buff coloured eggs, about 3 mm across, wrapping each in vegetation, such as the folded leaves of aquatic plants, or within beds of blanket weed. This provides some protection from predators. The eggs are different in size and colour from those of the great crested newt, but are indistinguishable in the field from palmate newt eggs.

The tadpoles hatch after about two weeks, remaining attached to water plants for a few days before becoming free swimming. They feed on small invertebrates such as water fleas. They are a pale brown in colour with a lighter belly and small dark spots. The gills are feathery and backward facing and the pointed tailfin has no filament. They spend most of the time hidden in vegetation or in detritus at the bottom of the pond, and are heavily preyed on by beetles, dragonfly nymphs, fish and larger newts. Between July and September the gills are absorbed and the metamorphs leave the pond, although late hatching or slow-growing larvae may remain in the pond over winter to emerge the

Smooth newt egg   *Metamorph just leaving the pond*

following spring. They do not return to water until they are ready to breed at two or three years old, spending the time on land hiding in crevices or under leaf litter, often clustered together under logs or stones. Smooth newts feed on a wide variety of prey including pond shrimps and water slaters *Asellus sp.* On land they have a similar diet to other newts, feeding on a range of small terrestrial invertebrates.

Habitat

The smooth newt regularly occurs in ponds, including garden ponds, pools and ditches containing a range of submerged and emergent vegetation. It avoids deep, heavily shaded and acidic water, preferring neutral conditions and hard water. In larger pools or lakes which support fish it can be found in low numbers around the shallow vegetated margins. As with the other newts, surrounding habitat needs to provide foraging and hibernation sites as well as dispersal corridors to other ponds.

In late summer and autumn they are often found under logs, stone paving slabs and in dry stone walls which provide plenty of cracks and crevices. Numbers in the countryside have decreased because of loss of suitable habitat, but garden ponds are a welcome new resource for this species.

Status, Threats and Legal Protection (2006)

The smooth newt is common within Britain. Numbers have declined both nationally and locally due to pond losses. Other threats include the unregulated introductions of fish into newt breeding ponds.

It is protected under Section 9 of the Wildlife and Countryside Act 1981 in relation to sale only.

National and Local Distribution

The smooth newt is distributed across much of Britain and is also found in Ireland. As it cannot tolerate very acidic water it is absent from many upland areas, but elsewhere it has a wide range of habitat preferences. Studies have shown that it has very similar habitat requirements to the great crested newt, choosing to breed in well vegetated seasonal ponds. However because of its shorter breeding cycle it can also breed in smaller ponds where the great crested newt would fail to breed successfully due to dessication. Recent pond surveys have indicated that smooth newt numbers decrease where thriving populations of great crested newts exist due to great crested newt predation on the small smooth newt tadpoles.

Smooth Newt
Triturus vulgaris

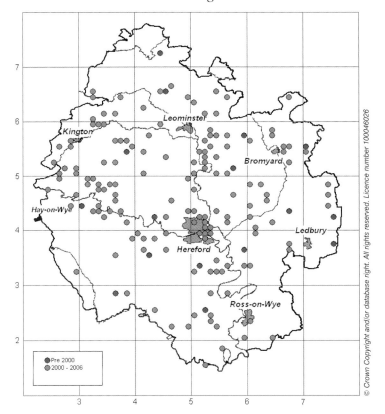

The smooth newt occurs throughout Herefordshire except in the low nutrient upland ponds which are rather scarce within the county. It is the most common newt because of its broad preferences and is the most likely one to occur in urban garden ponds. Large populations have been recorded in well-maintained garden ponds and in farm ponds lacking great crested newts.

During the HPN Project larval surveys were carried out. These could not distinguish between smooth and palmate newt tadpoles, resulting in under-recording of both these newt species relative to great crested newts.

Note

It has been proposed that the scientific name of the smooth newt be revised to *Lissotriton vulgaris*. (Frost *et al*. 2006.)

Palmate Newt *Triturus helveticus*

Male palmate newt in breeding condition. The filament on the end of the tail is very obvious

Description

The palmate newt is the smallest British newt, only about 6.5 cm long. The male is usually smaller than the female. Both male and female are a dull olive-brown above with black stippling. They have few, if any, spots on the underbelly and no spots at all on the pale throat, and both have a pale bar above the hind legs. The male also has a dark eye stripe which extends to the neck, and, on the tail, two rows of dark spots with a toffee-coloured strip running down the middle.

In the breeding season the male can easily be distinguished from the males of the other newt species by its black webbed hind feet, the low smooth crest along the back and tail and the hair-like filament on the end of the tail.

Webbed hind feet of the breeding male

The female closely resembles the female smooth newt, the principal difference being tubercles on the hind feet of the palmate newt. The typically plain throat and underbelly contrast with the dark markings on the underside of the smooth newt.

Life cycle

Palmate newts generally over-winter in hibernacula on land, migrating to their breeding pools in early spring. They spend the day hidden in vegetation in the pond, becoming more active at dusk when they move to more open areas. Once in the

Phyl King

Female palmate newt showing pale bar above the hind leg

pond the males perform elaborate courtship displays similar to those of the smooth newt, resulting in the fertilisation of the female's eggs. The female lays 200 to 300 buff coloured eggs on small underwater leaves, folding each leaf over in the same way as the smooth newt so that in the field it is impossible to distinguish the eggs of the two species. They are laid from March to June, but tadpoles that hatch late in the season will have to over-winter and may not survive. The eggs are heavily predated, particularly by great diving beetle larvae, and less than 20% of eggs laid may actually hatch. Photos of the similar egg and metamorph of the smooth newt are on page 32.

The palmate newt is locally considered to be the most aquatic of the three newts sometimes taking the opportunity to breed in ponds in July, although many adults will have left the pond by then. In some populations we have seen large numbers remain in the water throughout the summer and early autumn and even into winter.

Metamorphs leave the pond about three months after hatching, not returning to water until they are ready to breed at two or three years old.

Palmate newts eat whatever suitable sized prey is available. On land they feed on a range of invertebrates including springtails and soil mites.

Habitat

The palmate newt is generally found in more acidic ponds than the other two newt species and is associated with softer water habitats. However, there is a high degree of habitat overlap with the smooth and, to a lesser extent, with the great crested newt. The palmate newt can be found in a range of still water habitats, typically seasonal ponds without fish, although it can occur in quite large numbers around the shallows of large, well vegetated pools which do support fish. It is also found in very small water bodies such as tyre ruts on forestry tracks where it can be the only amphibian species. Where it occurs it is often very plentiful. As with the other newts the surrounding habitat needs to provide foraging and hibernation sites as well as dispersal corridors to other ponds.

Phyl King

Status, Threats and Legal Protection (2006)

The palmate newt is common within Britain. However, numbers have decreased as a result of land drainage, infilling of ponds and habitat neglect. Other threats include the uncontrolled spread of fish into breeding ponds. It is protected under the Wildlife and Countryside Act 1981 in relation to sale only.

National and Local Distribution

Distribution is patchy throughout Britain. It is the newt species more frequently associated with upland areas, occurring in tarns and lochans, and is the only one found in mountainous parts of Scotland and Wales. It is

Female palmate newt showing pale unspotted throat and yellowish underbelly

less frequently found in ponds with high natural alkalinity on the North and South Downs and in ponds on the Cotswold escarpment. The palmate newt is scarce in, or totally absent from, large parts of the country, particularly in the east where soft water habitat is less abundant.

In Herefordshire the geological conditions seem to be ideal for palmate newts. They are found throughout the county with the exception of the south-east around the Marcles. They are particularly suited to ponds on the Devonian sandstone which covers most of the county, and their distribution ends abruptly where the county meets the clayland landscapes of Gloucestershire and Worcestershire. Larger populations seem to develop in ponds on the higher ground where the Devonian St. Maughan's formation, which contains a greater proportion of sand, is more prevalent. These ponds are generally more acidic. Fewer records have come from ponds in the low lying areas of the Herefordshire plain where there is more clay and alluvium.

Significant populations have been recorded at British Camp reservoir on the Herefordshire/Worcestershire border, elsewhere on the Malvern Hills, and also to the west of the county. Palmate newts have been found over-wintering in ponds in surprisingly large numbers: over 300 were recorded at Byton Church pond during restoration work in February 2005, nearly 60 in King's Thorn in January and February 2001 and 38 in a Hereford school pond in January 2006.

Note

It has been proposed that the scientific name of the palmate newt be revised to *Lissotriton helveticus*. (Frost *et al.* 2006.)

Palmate Newt
Triturus helveticus

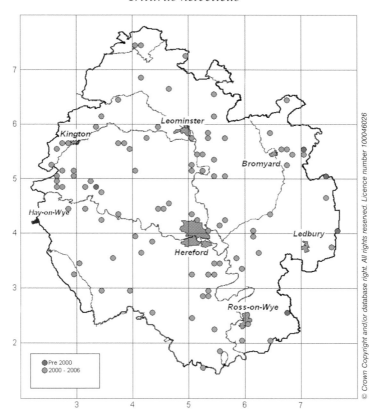

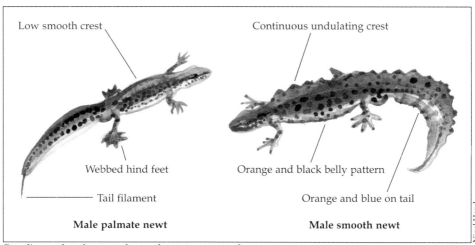

Low smooth crest	Continuous undulating crest
Webbed hind feet	Orange and black belly pattern
Tail filament	Orange and blue on tail
Male palmate newt	**Male smooth newt**

Breeding male palmate and smooth newts compared

Nigel Hand

Common Frog *Rana temporaria*

Common frog

Nigel Hand

Description

The common frog is the most frequently encountered amphibian. Its length ranges from 6 to 10 cm, the male being smaller than the female. It has a smooth, moist skin which is very variable in colour, the background usually ranging from olive, yellow or brown to grey. Some females are russet in colour with small bumps under the skin. The back, head and flanks have a very variable pattern of brown or black blotches, some individuals even lacking these blotches completely. The majority have stripes or bars across their legs.

Albino common frog

Will Watson

Distinctive colour patterns are noticed in some populations, which may in part be due to the frog's ability to vary its skin colour according to environmental conditions. Albino frogs with yellowish skin and red eyes occur from time to time. Two of these have been recorded in Ledbury, in 2004 and 2005.

Frogs have prominent eyes with a speckled yellowish/brown iris and slightly oval pupil. They also have highly visible circular eardrums located behind the eyes and surrounded by a triangular dark patch. They have longer back legs than toads making them strong swimmers and enabling them to jump six times their own body length when disturbed. The front legs are much shorter and in males have a thickened white pad on the first finger. These nuptial pads become dark and roughened in the breeding season and enable the male to hang on to the female during mating.

Life cycle

Common frogs mostly spend the winter in hibernacula on land, usually under piles of stones or logs, sometimes in the compost heap, although many males hibernate in the silt at the bottom of the pond, absorbing oxygen from the water through their skin. When the pond is iced over for long periods this may lead to frogs suffocating, so it is important to maintain a hole in the ice in garden ponds. Amazingly frogs can survive being frozen for a short time.

In February or March on mild wet nights frogs emerge from hibernation and migrate to their breeding ponds, where males congregate in large numbers waiting for the arrival of the females. The low purring croak of large numbers of male frogs is a sound well known to owners of even very small garden ponds.

Once the females arrive the males struggle and jostle with each other to grab a female which they then cling to in an embrace known as amplexus. They remain together until the female spawns and the clump of 1000 to 2000 eggs is fertilised by the male. The round, black eggs are surrounded by transparent jelly which absorbs water and swells to four times the original size, providing protection against predators. Usually frogs spawn within a few days of each other,

A pair of frogs on the way to the pond and wth freshly laid spawn; tadpole with hind legs developing; froglet almost ready to leave the water with the tail not yet absorbed

Phyl King

Great crested newt feeding in frog spawn

Phyl King

often in the warmer shallow edges of the pond, resulting in a seething mass of activity, but spawning may be spread over several weeks especially if interrupted by freezing temperatures. In ideal conditions the pond may be blanketed with spawn. After spawning most of the adults leave the pond and seek refuge on land. They emerge on damp nights to feed on slugs, snails, beetles, earthworms and flies, so should be a welcome resident in any garden.

On hatching the tiny black tadpoles remain in the disintegrating jelly for a few days before dispersing, when groups of them can often be found in sunny spots grazing on algae. As they grow they spend more time at the bottom of the pond and their skin becomes a speckled greenish brown in contrast to toad tadpoles which remain jet black. Unlike toad tadpoles frog tadpoles do not have a toxic skin and are eaten by a large number of aquatic species such as dragonfly nymphs, water beetles and fish. At night newts can often be seen amongst the spawn preying on eggs and tadpoles.

By May or June tadpoles complete their development into froglets, and hundreds may be seen leaving the water. At this stage they are very vulnerable and are heavily preyed on by many species, including herons, crows, blackbirds, grass snakes, foxes and hedgehogs. They immediately seek shelter under decaying vegetation, in holes in the ground or under logs. Later in the summer in dry conditions many will become inactive. In autumn they find a permanent site for hibernation. They become sexually mature at two or three years old, and may return to the pond annually to breed.

Habitat

Frogs can breed in a wide variety of water bodies from very small ponds to the shallow edges of large lakes. Occasionally spawn is even found in tractor ruts in muddy fields,

Nigel Hand

Frog spawn laid in a tractor rut

Phyl King

The common frog spends most of its life on dry land

A typical garden pond with a good population of common frog

although breeding in such sites tends to be unsuccessful due to desiccation. They occasionally spawn in slack pools in slow running water. The best ponds for frogs are shallow, groundwater-fed, preferably covering a large area. Scrapes and hollows in marshland are ideal. Large populations may also develop in garden ponds particularly where there is not undue predation from newts.

Good vegetation structure is not essential for frogs, which often thrive in ponds dominated by algae. Unlike newts, frogs do not seem to avoid ponds with fish so can be found spawning in larger pools and lakes, although at lower densities. They spend very little time in the water and need damp terrestrial habitat for shelter and feeding, and frost-free places to hibernate. Habitats include rough grassland, gardens, hedges and woods.

Status, Threats and Legal Protection (2006)

The common frog is widespread in Britain. It is protected under Section 9 of the Wildlife and Countryside Act 1981 in relation to sale only.

National and Local Distribution

The common frog is the most widely distributed of our native amphibians. It is found in all counties in Britain and Ireland, is widespread throughout most of Europe, and is found across Asia to Japan.

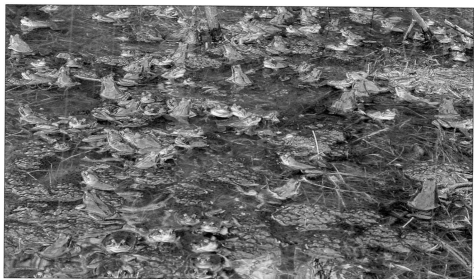

Tupsley Quarry which has a large breeding population of common frog

Common Frog
Rana temporaria

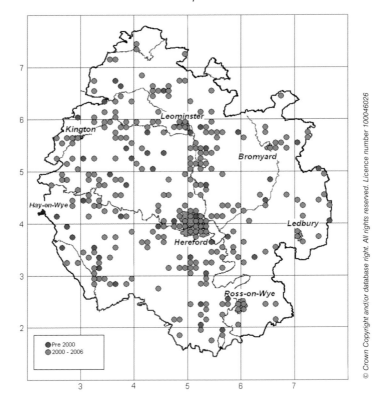

Frogs are very adaptable and are able to colonise a variety of still water habitats. They can tolerate a wide range of environmental conditions, breeding in ponds which are acidic and low in nutrients as well as those with high alkalinity, and can be found at both high and low altitudes. In spite of their widespread distribution, evidence suggests that there has been a decline in frog numbers on farmland. However, this has been partly compensated by the successful colonisation of garden ponds. It has been estimated that one out of five gardens now has a pond.

The common frog's distribution in Herefordshire is typical of the national situation; it is widespread across the whole county as well as being the most common amphibian. The highest density of records comes from garden ponds, with urban areas becoming particularly important.

We have records from most farmland ponds surveyed but breeding numbers are generally low. Excellent breeding sites seem to be rare; sites which stand out include Tupsley Quarry at Hampton in Hereford where in 2003 an estimated 7000 breeding individuals were present in this relatively small area, and a garden in King's Thorn where 113 adults were found hibernating at the bottom of a pond in 2001.

Common Toad *Bufo bufo*

Simon Williams

Common toads in amplexus. The female is much larger than the male. The copper coloured eye and horizontal pupil are very obvious.

Description

The common toad is much stockier in appearance than the common frog and has a warty skin, which can become dry, unlike the smooth skin of the frog. The upper surface and flanks vary from greyish brown to a dull green colour, the female often more ruddy. It frequently has dark blotches which, as in the frog, vary in size and intensity of colour. Very occasionally common toads may be more exotically coloured; orange individuals with black spots have been noted. Like other amphibians it can lighten or darken its skin tone to suit its environment. In Herefordshire brick red toads are occasionally encountered; this may serve as camouflage against the locally occurring sandy red soil. The toad has two half-moon shaped paratoid glands behind the eyes, unlike the frog which has a circular eardrum behind the eye. The eye is a very striking copper colour with a horizontal pupil.

Nigel Hand

Variations in toad colour

Ball of male toads with a single female

Pair of toads in amplexus

Pair with long strings of spawn

The raised warts on the skin and paratoid glands contain toxins which are secreted when the toad is attacked, providing an effective deterrent to predators. A dog that picks up a toad will salivate and froth at the mouth and while it is distracted the toad will make its escape. The hind legs are shorter than those of the frog, so instead of the long leaps of the frog the toad moves in a series of short hops or a crawl.

The male measures up to 6 cm, the female considerably larger at up to 9 cm. During the breeding season the male can also be distinguished from the female by the presence of dark-coloured nuptial pads on the inner three fingers. Additionally when picked up it will squeak, unlike the female which does not vocalize when disturbed. Toads can be very long lived, some reaching 30 years in captivity, although they would be lucky to reach ten in the wild.

Life cycle

Toads spend only a short time in water, just a week or so in spring for breeding. On damp mild nights, usually in late February and March when the temperature reaches 7°C or above, they make their

Jet black toad tadpole contrasts with the speckled frog tadpole

Phyl King

Toadlet ready to leave the water

way from winter hibernation to their traditional spawning ponds. They tend to move within a few days of each other and it is during these mass migrations that many are killed on roads which cross their traditional routes. Males can be particularly vulnerable as they may congregate on a road, possibly as a good vantage point from which to observe and intercept a female.

There is a male bias in toad populations, leading to great competition for females at the pond, and many females arrive already in amplexus. Single males struggle to supplant ones already paired, and balls of males can form round a single female in a mating frenzy during which the female is occasionally drowned. The males in particular are very vocal during breeding and emit a high pitched call which is very different from the purring call of the male frog. Large males have louder and slightly deeper calls than smaller individuals, an indication of their size, and hence ability to ward off rivals.

In contrast to the familiar clumps of spawn laid by frogs, toad eggs are produced in long strings of jelly up to 250 cm long that are wound round submerged plants. Common toads spawn in deeper water and normally later than frogs. However they occasionally spawn at the same time as was seen at Hampton Court pool, Hope-under-Dinmore in March 2005, when, following a cold spell of weather, large numbers of both species were observed in courtship.

Most adults leave the water after breeding, and can move up to 1.5 km to find suitable shelter, usually under logs or piles of stones, or in small mammal holes. On mild damp evenings they come out to feed on a wide range of invertebrates and will snap at anything that moves, which includes slugs and insect larvae, so they are a welcome ally of the gardener. They are solitary whilst on land. They hibernate in holes in dry banks, disused mammal burrows or even cellars.

The tadpoles hatch after about a fortnight. They are jet black and protected from predators by toxins in their skin. As a result smooth and palmate newts and many fish find them unpalatable. However they are eaten by great crested newts and the larvae of dragonflies and other aquatic insects. They gather in shoals in warm areas of the pond and so are much easier to observe than frog tadpoles. In June or July, usually after rain, the tiny toadlets emerge from the water and make their way into the cover of long vegetation. They will not return to water until they reach maturity, which in males is between two and three years and females three and four years.

In spite of protection from skin toxins, toads have a range of predators including otters, mink and polecats, which will first skin them or turn them onto their backs before eating them to avoid the toxins. They are also eaten by grass snakes, hedgehogs, buzzards, corvids and herons.

Habitat

Toads have more specific breeding habitat requirements than frogs. They prefer larger ponds, 1000 m^2 being the estimated optimal size, and tend to avoid small ones, so rarely breed in small garden ponds. Breeding ponds often support fish. Our surveys have indicated that large breeding populations are associated with pools which have a good vegetation structure. Those with poor vegetation structure support only low breeding populations regardless of their size.

Toads wander considerable distances from their breeding ponds and are frequently found in suitable gardens which provide shelter, damp areas to protect their skin from dessication, and opportunities to forage. They may take up residence under an upturned flower pot or find refuge in a greenhouse.

Common toads can tolerate drier habitats than frogs. In open countryside woodlands in particular seem to suit their requirements, and here their skin is excellently camouflaged in the leaf-litter. Scrub and rough grassland also provide suitable conditions. They have been recorded returning to the same site year after year.

Phyl King

Toad well camouflaged in leaf litter

Status, Threats and Legal Protection (2006)

The common toad is widespread in Britain, but questionnaire surveys have indicated that numbers are declining in central, eastern and south-eastern England. The cause of this decline is unknown, but concern has been expressed both nationally and locally about the impacts of road mortalities. On the minor road leading into Bodenham village from Dinmore Hill, HART members have counted more dead toads than live, with upwards of 200 dead animals being seen on a single night. Pollution from agricultural run-off and other diffuse forms of pollution have also been cited as a cause of decline.

The common toad is protected under Section 9 of the Wildlife and Countryside Act 1981 in relation to sale only.

Common Toad
Bufo bufo

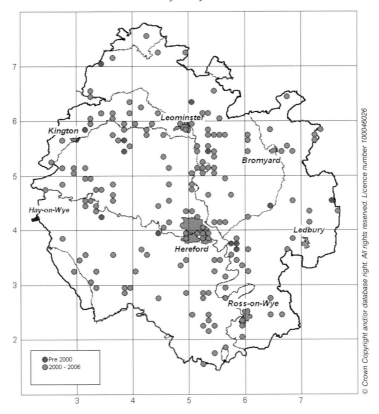

National and Local Distribution

The common toad is widely distributed throughout Britain although it is absent from Ireland. It is not as widely distributed as the common frog in upland areas.

In Herefordshire, as with the rest of the country, common toads are regularly associated with larger water bodies and are absent from many of the smaller ponds. Large pools in which toads breed are relatively well represented in the county. Fishing pools, many of recent construction, are popular breeding sites, as are irrigation reservoirs, old quarry workings and mill pools. Many thousands of toads have been seen migrating to Bodenham Lake, a restored sand and gravel working. Large numbers of our records are of sightings on land, many being collected on the county's roads during the spring migration. Aggregations of toads are seen near to the breeding pools and these are reflected in the records.

Viviparous or Common Lizard *Lacerta vivipara*

Viviparous lizard

Nigel Hand

Description

The viviparous, or common lizard as it is also known, has short legs, a rounded snout and a thick neck in relation to its body. The adult length from snout to tail tip is 10 – 15 cm, the longest recorded in the UK being 17 cm. The colour and patterning varies from animal to animal: the majority are brown with the occasional green individual being seen. This background colour is broken up by light and dark coloured spots known as ocelli which are usually more pronounced on the male. Both sexes generally have a vertebral stripe which tends to be more pronounced in the female. The underside of the male is orange with dark spots, whereas that of the female is cream. The female has darker flanks than the male and the tail is approximately the length of the body. In the male the tail is longer than the body and is thicker at the base relative to that of the female. The young are black or bronze. Melanistic viviparous lizards occur but there are no records in Herefordshire.

Green variant of the viviparous lizard

Nigel Hand

Life cycle

The viviparous lizard emerges from hibernation in early spring, from February to March, and courtship and mating occur

Contrasting belly colours of male (left) and female (right)

Nigel Hand

soon afterwards. First the males shed their skins and may fight to establish dominance. Mating involves the male taking hold of the female in his jaws. Once mated the females need to increase their basking times to aid development of their embryos over a period of up to three months.

Between three and ten young are born in July, or later if temperatures are cooler than average. There is some evidence to suggest that female viviparous lizards need their winter hibernation temperature to fall below 10°C for a few weeks to spur ovarian development. This species is extremely cold tolerant and has a more northerly distribution than any other European lizard. Basking is important and they have favoured basking spots, usually fallen branches or clumps of vegetation, which are used daily. Particularly favourable basking sites may attract groups of lizards. When basking the lizard flattens its body to absorb heat over the greatest surface area possible.

Young are born live and measure 37 – 44 mm at birth. The mortality rate in the first year is as high as 90%, but individuals may live as long as 12 years.

Viviparous lizards' prey consists of spiders, grasshoppers, butterflies, moths and aphids.

Young viviparous lizard

Nigel Hand

Flattened bodies of basking viviparous lizards

They are preyed upon by birds, including buzzards, kestrels, corvids, pheasants and blackbirds as well as shrews, stoats, weasels, hedgehogs, cats, young adders and the common toad. As a defence mechanism they have the ability to shed their tail, which regrows.

Habitat

Ideal habitat includes unmanaged, south facing, damp, tussocky grassland; scrub covered hillsides or banks; heathland and woodland rides. The majority of recorded sightings in Herefordshire have been on woodland rides and scrub covered hillsides.

Status, Threats and Legal Protection (2006)

Under the 1981 Wildlife and Countryside Act the viviparous lizard is protected from intentional killing, injury or sale.

National and local distribution

The viviparous lizard is one of the most widely distributed vertebrates in the world, its range extending from Britain and Ireland, across northern and central Europe and Asia to the coast of the Pacific. However, in Herefordshire like the adder it is very localised, and definitely not 'common'. Agricultural intensification, overgrazing and the wide use of chemicals are all possible factors contributing to its limited distribution. Where colonies are found their population numbers can be high. They are recorded in the north of the county at Bircher, Croft, Checkley, the Malverns, Bromyard Downs and the Welsh border.

Gerald Leighton (1903) writes: 'I am convinced from some years' observation that the

Typical lizard habitat   *Lizard site at Croft Ambrey*

Viviparous Lizard
Lacerta vivipara

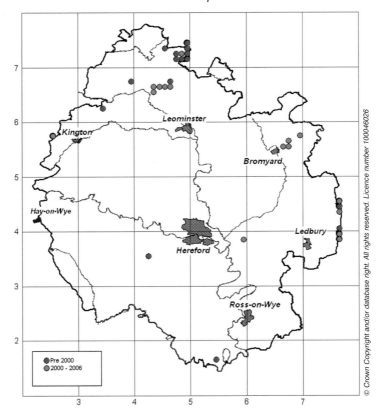

viviparous lizard is very rare in South Herefordshire while the slow-worm may be found in every old quarry.' This statement remains true for the whole county a hundred years later.

Like the adder, the viviparous lizard's dispersal abilities are limited, and colonisation of new habitat can only occur if this is close to an existing population. However, unlike the adder, when the conditions on a new site are suitable, a viviparous lizard population may expand very quickly.

Interestingly two records exist for the sand lizard *Lacerta agilis* in or near the county. A sand lizard was recorded in 1861 on the Doward; and the Worcestershire Naturalists' Club (1898-99) reported that a sand lizard was caught on May Hill during a field trip on 26th July 1898. Leighton (1903) mentions sightings in Worcestershire, Tenbury, and the Wyre Forest, but he never saw any specimens and did not obtain conclusive evidence of identification. He does suggest the possibility of misidentification as there are many colour variants of the viviparous lizard, but maybe the sand lizard's range was very scattered and sporadic. We may never know.

Note

It has been proposed that the viviparous lizard should be in its own genus *Zootoca* but this has yet to be universally accepted.

Slow-worm *Anguis fragilis*

Phyl King

Female slow-worm

Description

The slow-worm is a legless lizard but is regularly mistaken for a snake. The adult is usually 30 - 38 cm in length, the largest recorded being 48.9 cm. It has a polished, cylindrical appearance and, unlike a snake, has eyelids. The tongue is short, broad and flat. Many slow-worms have regenerated tails.

Male and female tend to differ in colour. The female's overall colour ranges from copper or reddish brown to mahogany, with darker flanks, usually brown or black. It has a thin dark vertebral stripe and a small head. The colour of the male is more uniform, varying from grey to brown, and it only very occasionally has a vertebral stripe. Its overall build is thicker than that of the female, and it has a larger head. Some males, including those in Herefordshire, have blue spots. Melanistic or albino slow-worms are rare.

Beneath the scales is a layer of bony plates called osteoderms. These plates enable the slow-worm to burrow through soil and vegetation more easily, but their presence means that it does not have the supple feel of the snake.

Life cycle

Emergence from hibernation takes place during March, males emerging first. Mating is later than in the other viviparous reptiles, usually in May or June. During this period males will fight, writhing and biting each other hard enough to cause scarring. Once a female is located a male will hold her neck region in his jaws while copulating, which can

Nigel Hand

Above left: Mating slow-worms. The male is holding the female's head in his jaws

Above right: Hatching young

Left: Female slow-worm with newborn

last for hours. Like adders they tend to breed every other year. The young are born covered in a membrane, the pellicle, from which they shortly struggle free. Between five and twenty-five babies are born, usually from August to September, although in a cool, damp summer it may be later. At birth they are 7 – 10 cm long, their length doubling within the first year. Newborn slow-worms are very attractive: gold or yellow with a dark vertebral stripe.

Slow-worms live primarily underground. They seldom bask in the open, but may do so among vegetation. They often raise their body temperatures by moving under objects that are warmed by the sun, especially discarded sheets of metal, such as corrugated iron, or rubbish. On good sites large numbers can be found by looking under such refugia. In Haugh Wood slow-worms are regularly seen thermoregulating in wood ant nest

Will Watson

Slow-worms under refugia

Slow-worm eating an earthworm

piles. They are more active on warm, showery, overcast days, possibly because these conditions bring out their main prey, slugs. The white slug *Deroceras reticulatum* appears to be favoured over any other slug, and they also take earthworms and spiders. They follow their prey for a short time before striking, gripping and swallowing it.

Predators are mainly birds of prey, pheasants, corvids, foxes, badgers, hedgehogs, rats and very occasionally adders. Their defence strategies include shedding the tail, which will grow back, or defecating on handling. In general they do not bite in self defence.

Habitat

Slow-worms occur in woodland rides, rough unmanaged grassland, hedgerows, banks, railway embankments, heathland, brown-field sites, and suburban gardens where they favour compost heaps. The more overgrown areas of churchyards also provide an important habitat.

Status, Threats and Legal Protection (2006)

Slow-worms are protected under the 1981 Wildlife and Countryside Act, making it an offence to intentionally kill or injure them, or sell them without a licence.

Typical open woodland habitat

A churchyard with a good population of slow-worms

Slow-Worm
Anguis fragilis

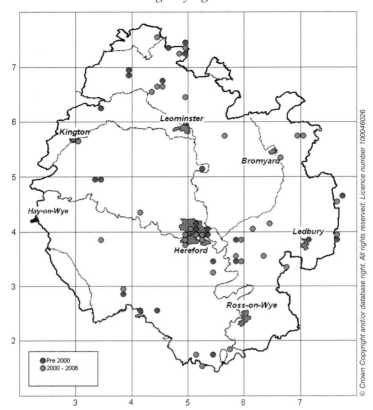

National and Local Distribution

Slow-worms occur across England, Wales and Scotland, sometimes in very large concentrations. Population densities can be high; as many as 600 – 2100 per hectare have been recorded in Dorset and Kent. Slow-worms are under-recorded in Herefordshire. It is likely that they are ubiquitous, although in low numbers, throughout the county. Records show that heavily populated areas include Fownhope, Woolhope, Knighton, Bircher, Croft, the Mortimer Forest, Checkley and the Malverns.

Gerald Leighton (1903) says of Herefordshire: 'In some parts of the county the slow-worm is very abundant ... In the Monnow Valley the slow-worm is the only lizard I have ever seen, and the same remark applies to Kentchurch and Ewyas Harold.'

In Hereford City they have been recorded on the Holmer trading estate, Rotherwas Industrial Park, at Belmont and along the railway embankments, which provide a good corridor for populations to move through.

With more comprehensive recording there would probably be a considerable increase in the number of populations identified. More records please!

Grass or Ringed Snake *Natrix natrix*

Nigel Hand

Large female grass snake

Description

The grass snake is Britain's largest snake with females averaging 75 to 80 cm in length and males 65 to 75 cm. One of the largest specimens ever recorded in Britain, measuring 177.5 cm, was killed in South Wales in 1887 (Leighton 1901); the largest recorded by the county reptile recorder was found in Ewyas Harold in 2004 and measured 112 cm.

It is a slender, fast moving snake. The body is olive green to grey with a yellow and black, or orange and black, collar. Occasionally this collar colour is absent in older, larger specimens. The belly has a black and white chequered pattern with vertical black stripes on the flanks of some snakes. Albino and melanistic grass snakes do occur but these are rare.

Fergus Henderson

Dark variant of the grass snake

Life cycle

Grass snakes spend the winter in hibernacula in hedgerows, hillsides and banks, usually amongst tree roots or in mammal burrows. They emerge from hibernation in March to April and occasionally, at some sites, may be seen basking in groups with adders or slow-worms. Mating takes place from April to May. A group of males will trail a female, locating her by the pheromone she secretes. All of the males will attempt to mate with her

Grass snake lying on its back with belly exposed

occassionally forming a ball of writhing snakes. There are records of this phenomenon in Herefordshire. Only one male will mate with the female.

In the summer months, mating complete, they disperse to their feeding areas. These tend to be damper places where they find their prey, which consists mainly of frogs, toads, newts, fish, small rodents and birds. Grass snakes are active hunters and when prey is caught it is swallowed alive, usually head first. This snake species is wide ranging, especially females searching for suitable egg laying sites. Between June and August they may travel up to a

Grass snake head: note the round pupil

Grass snake feigning death

couple of kilometres to find the ideal spot, such as a large heap of rotting vegetation, or pile of manure. One extremely good egg laying site was an active saw mill where large numbers of eggs were laid in the wood chip piles. An especially large compost heap may attract a number of females. Each female lays between 5 and 40 leathery-shelled eggs, 15x20 mm in size, which can adhere to each other to form a clump. Incubation lasts between 40 and 70 days depending on factors such as compost temperature.

Garden habitat for grass snakes

Between August and October the 15 - 20 cm long hatchlings emerge, shed their skins (slough) and disperse to begin hibernation in October. Male snakes are sexually mature at about three years old when at least 50 cm long; females at four to five years and 60 cm. Grass snakes regularly slough their skins. Females will shed prior to egg laying, and a large snake skin on the garden compost heap is a good indicator of a female's presence.

Predators of grass snakes include buzzards, corvids, foxes and cats. Many female grass snakes are killed crossing roads in search of favourable egg laying sites.

Given the chance, grass snakes can live up to 15 years. They are harmless, and if molested may hiss, thrash around and strike, but very rarely bite. Older females tend to behave in this way but many specimens when handled will 'play dead', going limp with the mouth open and tongue out. If placed on the ground they will lie on their back and when turned over roll on to their back again. As soon as the possible predator loses interest and moves away the snake will slip into the undergrowth. They may also release a foul smelling liquid from the anal gland.

Habitat

Grass snakes occur in woodland rides and hedgerows, on commons and wetlands, by ponds, lakes and rivers, in gardens, parks and allotments, especially those bordering wild areas, and on farmland.

These snakes are excellent swimmers and many sightings are from garden ponds or lakes, even the river Wye. They are occasionally referred to as the water snake.

Status, Threats and Legal Protection (2006)

Grass snakes are protected under the 1981 Wildlife and Countryside Act, making it an offence to intentionally kill or injure them, or sell them without a licence.

National and Local Distribution

Grass snakes are found in England and Wales. They are concentrated in the southern counties with sparse distribution in the north. In Herefordshire they are widespread with records in the South Wye region, Ross-on-Wye, Ewyas Harold, Stoke Edith, Fownhope, Bircher Common and the border with Shropshire, and the city outskirts of Holmer and Rotherwas. The piles of hop waste at Ashperton, Tarrington and Stoke Edith have provided good egg incubation sites, but sadly hop growing is diminishing in the county.

Grass Snake
Natrix natrix

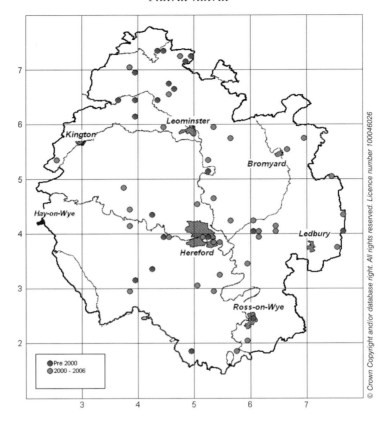

HART responds to snake sightings in gardens in Herefordshire. Grass snakes are the most frequently reported reptile along with slow-worms, but both are often misidentified as adders. Gardens with ponds and open compost heaps prove attractive to this species, but sealed plastic compost bins are not grass snake or slow-worm friendly. Pea netting in vegetable allotments or spread over fish ponds to deter cats or herons can be a hazard as grass snakes can get stuck in the mesh.

The grass snake is under recorded in Herefordshire and more records are needed.

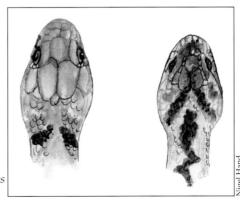

Grass snake (left) and adder (right) head markings compared

Nigel Hand

Adder or Northern Viper *Vipera berus*

Male adder just after the spring slough

Description

The adder is the only venomous snake found in the British Isles. It is a relatively small snake with a characteristic zigzag pattern along its back. The average length of the male is 50 – 60 cm and the female 55 – 65 cm. Gerald Leighton (1901) recorded one of 28.5 inches (72 cm) found in the Monnow Valley in the late 1890s. His data show that the average length of the adult male was 24 inches (61 cm) and the female 25.5 inches (65 cm). This was, he claimed, 'being much above the average of most localities'.

The male and female can be distinguished by their colour, especially just after shedding, or sloughing, their skin. The male's background colour is grey to silvery white with a black or dark brown zigzag, whereas the female's is brick red to brown with a dark brown zigzag. The female tends to be stouter than the male. A 'v' or 'x' mark behind the head is present in both sexes. Adders have an elliptical pupil, unlike the round pupil of the grass snake. Melanistic or 'black adders' do occur and even on these the zigzag can be discernable. Leighton (1901) mentions seeing black to light brown and white adders in the Monnow Valley.

Life cycle

The males are the first to emerge, from late February, once the temperature steadies above 10°C. After basking near the hibernacula they disperse to await the emergence of the females, two to six weeks later. Adders need to bask to condition themselves for the rigours of combat and mating.

By April to May the males have shed their skins and now wrestle any other males in the vicinity of a female adder in a contest of strength. This is known as 'The Dance of the Adder' and is spectacular to see, as the males chase each other through the undergrowth, twisting, twirling and rearing up against each other in an effort to push the opponent to the ground. The victor goes on to mate with the female, which he finds by detecting the pheromone she produces.

Mating complete, the partners stay together briefly before moving off separately to their summer feeding areas. Voles and mice are their main prey followed by amphibians, lizards and very occasionally birds.

Females are viviparous, giving birth to three to twelve live young in the vicinity of the hibernaculum between August and October. In 1899 and 1900 Leighton dissected 23 gravid female adders. Of these the average number of young was 13, but it ranged from 7 to 20. At birth the young are 14 – 18 cm long. Female adders usually breed every two years but this can extend to every three to four years. Males and females return to their hibernacula as the temperature drops in autumn.

Adders can fall prey to, buzzards, corvids, foxes and hedgehogs. They can live for over 15 years.

Nigel Hand

Newborn and young adders. Bottom right young gravid (pregnant) female

Nigel Hand

Male and female adders in courtship. The female is the stouter brown coloured snake

Nigel Hand

Gravid female adder

Male adder. Note how the pattern blends in with the bracken

Habitat

Adder habitats are scrub-covered hillsides, commons and heathlands, woodland rides and railway embankments especially with a south/south easterly aspect. Hibernacula can be under stands of bramble, gorse or bracken in old mammal burrows, which the adders often share with other species of

Typical adder habitat

Adder hibernaculum

Nigel Hand

Two females basking, showing considerable colour difference. The darker one may be much older

reptile and amphibian. These sites will be used every year and their existence is essential to the survival of an adder population. Populations can only expand to new sites when potential new habitat is within a km of an existing viable colony and even then population numbers may increase only very slowly, if at all (Sylvia Sheldon pers. com.).

Basking sites tend to be suntrap areas, protected from the wind and with escape routes to cover close by. Old ant mounds in unimproved grassland provide refuge and small stands of trees like birch, ash or elder provide protection from spring wind chill. Summer habitat tends to be in damper areas with an abundance of prey items.

A recent publication (Baker *et al*. 2004) states that habitat management is the factor most frequently regarded as impacting on adder and slow-worm populations. In spite of reports of individual sites being harmed, habitat management or creation was regarded as a positive factor at more than 40% of adder, and 50% of slow-worm sites.

Status, Threats and Legal Protection (2006)

The adder is protected under the Wildlife and Countryside Act 1981. It must not be intentionally killed or injured, or sold without a licence. As a venomous snake, it is also illegal to keep an adder in captivity, without the necessary licence, as required under the Dangerous Wild Animals Act 1976. In spite of this legal protection the adder has still been misguidedly seen as a potential threat on some sites in Herefordshire and populations have suffered irreversible damage. One site in Herefordshire was regularly burnt to rid it of adders!

National and Local Distribution

Adders occur throughout England, Scotland and Wales, with localised distribution in the Midlands. In Herefordshire most records for adder sightings are in the north of the county, the borders with Shropshire, Wales and Worcestershire, the Malverns and the Mortimer Forest.

In response to concerns about the status of the adder data gathered by national herpetological recorders showed that whilst some populations of adder were stable there was a national decline, with the Midlands being an area of particular concern (Baker, *et al*. 2004). A third of

Adder
Vipera berus

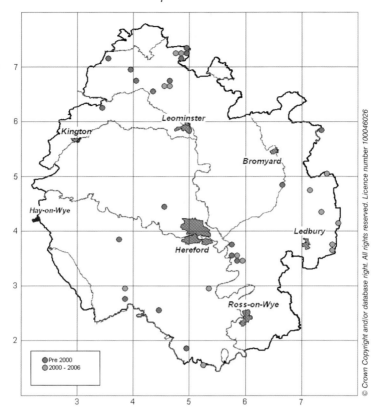

adder populations contained fewer than ten adults. Disturbance from public pressure was the most frequently reported detrimental factor. One third of sites were isolated and hence potentially at greater risk of local population extinction.

This is not just a recent problem as Gerald Leighton (1901) details this occurrence in the Monnow Valley over 100 years ago: 'The transition is sharp from arable or ploughed land to mountain and woodland. Obviously as civilisation advanced toward the valley from all sides, as the land became cleared and the plough succeeded the axe, the wilder animals which shun the approach of man would retire before him.' Leighton notes that an adder was killed at Credenhill Park. In recent years there have been no recorded sightings of adders on this site. He also states that the adder was a common sight to the south of the Wye, but that the grass snake was rare. However 100 years on the reverse seems to be the case.

Adder bites

Adders are not aggressive snakes and bites are extremely rare. If disturbed the snake will generally retreat to the nearest cover. Bites occur when adders are stepped on or threatened. The reaction to a bite can vary from minor swelling to severe allergic reaction and, in very rare cases, death. Medical advice should always be sought if bitten.

Notable adder behaviour dates in populations studied by Nigel Hand

Year	1st male emergence	1st female emergence	1st sloughed male	Interaction / Aggregations / Interesting behaviour
2004	*25th Feb*	*31st Mar*	*19th April*	*19th Apr* 5 males and sloughed skin in very close proximity. Combat, courtship and mating most likely around now.
2005	*4th Mar* 1 male	*21st Mar* 2 large females together	*11th Apr* male seen sloughing	*21st Mar* Male basking on a slow-worm *21st Apr* 6 males 2 females in close proximity. Combat, courtship and mating most likely around now.
2006	*1st Mar* 1 male	*29th Mar* Large female	*17th Apr*	*7th Apr* 3 males trailing each other, curling up together, also interesting head swaying. Large female 2 m away from this group. Another 3 males in close contact all basking within a metre of each other, roughly 10 m from the other group. A total of 14 adders seen, 3 females and 11 males. A pair of grass snakes was observed in courtship at same site. *26th Apr* Male and female in courtship. *28th Apr* 7 males 2 females in very close proximity. Males unsloughed.

Introduced species

Marsh Frog *Rana ridibunda*

A loud frog call was recorded in a pond at Madley in 2006 and identified by Will Watson as one of the water, or green, frogs. Analysis of the call by Dr. Julia Wycherley of Surrey Amphibian and Reptile Group indicated that it was likely to be the marsh frog, although analysis of the full mating call would be needed for complete certainty. However, 95% of all water frogs so far identified in the UK have turned out to be marsh frogs.

The marsh frog is native to the continent and all British populations originate from introductions. We do not know the source of the Madley marsh frog but we will continue to investigate the distribution and origin.

In 2004 Nigel Hand heard a green frog in a garden pond in Bodenham, but this was apparently an escapee from a neighbour who kept captive bred animals.

Wall Lizard *Podarcis muralis*

Wall lizards of Swiss and Italian origin, three males and two females, were introduced into the Mortimer Forest in 1981. Young were seen in 1985. They may still be lingering on but HART has no confirmed sightings.

Red-eared Terrapin *Trachemys scripta*

There have been a number of reports of terrapins in Castle Pool in the centre of Hereford. They have yet to be officially confirmed, but it is likely that they are red-eared terrapins, the species which was most commonly imported from America.

Bibliography

Arnold, H.R. 1995. Atlas of amphibians and reptiles in Britain. ITE research publication no.10. HMSO Publications.

Baker J., Suckling J. and Carey R. 2004. Status of the Adder Vipera berus and the slow-worm Anguis fragilis in England. English Nature Report number 546.

Beebee T. and Griffiths R. 2000. Amphibians and Reptiles A Natural History of the British Herpetofauna The New Naturalist Series, Harper Collins,

Beebee, T.J.C. 1985. Frogs and Toads. Whittet Books, Oxford.

Brandon, A. 1989. Geology of the country between Hereford and Leominster. Memoir of the British Geological Survey. Sheet 198. HMSO.

Bray, R. and Gent, T. 1997 Opportunities for amphibians and reptiles in the designed landscape. English Nature Science Series No. 30. English Nature.

Brian, A. and Harding, B. 1996. A Survey of Herefordshire Ponds and their value for Wildlife 1987-1991. Transactions of the Woolhope Naturalists' Club. Vol. XLVIII, Part 3.

Cooke, M.C. 1893. Our Reptiles and Batrachians. W.H. Allen, London.

Frazer, D. 1989. Reptiles and Amphibians in Britain. Collins New Naturalist Series. Bloomsbury Books. London.

Froglife. 2000. The Herpetofauna Workers Guide.

Frost, D.R. et al. 2006. The amphibian tree of life. Bulletin of the American Museum of Natural History 297: 1-370

Gent, A.H. and Gibson, S.D., editors 1998. Herpetofauna Workers' Manual. Joint Nature Conservation Committee. Peterborough.

Herefordshire Archaeology 2006. Sites and Monuments Record. www.smr.herefordshire.gov.uk

Holdsworth, E.W.H. 1863 Proceedings of the Zoological Society. Zoological Society of London.

Jackson, J.N. 1954. Historical Geography of Herefordshire from Saxon Times to the Act of Union, 1536 quoted in Herefordshire: Its national history, archaeology and history, Woolhope Naturalists' Field Club. S R Publishers Ltd, London.

Hilton-Brown, D. and Oldham, R.S. 1991. The Status of the widespread Amphibians and Reptiles in Britain, 1990, and changes during the 1990s. Nature Conservancy Council.

Leighton G. R. 1901 The Life History of British Serpents and their Local Distribution in the British Isles.

Leighton G. R. 1903. The Life History of British Lizards and their Local Distribution in the British Isles.

Lewis, C.A. and Richards, A.E. 2005. The Glaciation of Wales and adjacent areas. Logaston Press, Bristol.

Malmgren, J.C. and Bulow, P. 2001. Predator avoidance response by adult newts to fish cues. Evolutionary Ecology of Newts.

Mee, A. 1938. The King's England: Herefordshire. Hodder and Stoughton, London.

Morrison, P, 1994. Mammals, Reptiles and Amphibians of Britain and Europe. Macmillan.

Sheldon, S. 1988 – 2006 Annual reptile census and reports.

Worcestershire Naturalists' Club Transactions. 1898-99.

Recommended Reading

Amphibians and Reptiles A Natural History of the British Herpetofauna
Beebee T. and Griffiths R. 2000. The New Naturalist Series, HarperCollins,

Amphibians and Reptiles of Surrey
Wycherley, J and Antis, R. 2001. Surrey Wildlife Trust.

Dig a Pond for Dragonflies
British Dragonfly Society 1992.

Collins Guide to Freshwater Life
Fitter, R. and Manuel, R. 1986. Collins.

Great Crested Newt Conservation Handbook.
Langton, T.E.S., Beckett, C.L., and Foster, J.P. 2001. Froglife.

The Pond Book. A guide to the management and creation of ponds.
Williams et al. 2000. The Ponds Conservation Trust, Oxford.

Reptiles & Amphibians of Britain & Europe
Arnold, E.N. and Ovenden, D.W. 2002. Collins Field Guide.

Small Freshwater Creatures
Olsen, L.H., Sunesen, J. and Pederson, B.V. 2001. Oxford University Press.

The Wildlife Pond Handbook.
Bardsley, L. 2003. New Holland.

Shire Natural History Series:
The Adder; The British Lizards; The Common Toad; Newts of the British Isles.

English Nature Publications. Some of these can be downloaded from the internet.
Amphibians in your Garden. 2002. ISBN 1 85716 636 1
Farmland Wildlife - Past, Present and Future. 2004.
Great Crested Newts on Your Farm. 2003. ISBN 1 85716 698 1
Great Crested Newt Mitigation Guidelines 2001. ISBN 1 85716 568 3
Managing ponds for wildlife. 1998.
Old Meadows and Pastures. 2002. ISBN 1 85716 659 0
Reptiles in Your Garden 2003. ISBN 1 85716 711 2
Reptiles: Guidelines for developers 2004. ISBN 1 85716 807 0
Wildlife and Development. 2004.

Further sources of information

ARG UK www.arg-uk.org.uk

English Nature, now incorporated into Natural England. www.english-nature.org.uk

Froglife www.froglife.org

HART www.herefordhart.org

Herefordshire FWAG www.fwag.org.uk

Herpetological Conservation Trust (HCT) www.herpconstrust.org.uk

Herefordshire Biodiversity Partnership www.wildlifetrust.org.uk/hereford/biodiversity

Joint Nature Conservation Committee www.jncc.gov.uk

Natural England www.naturalengland.org.uk

Contributors to the Ponds and Newts Project

HART acknowledges with many thanks the following people who have contributed to the preparation of this book during the Herefordshire Ponds and Newts Project (2003-2006) through surveying and recording, survey team coordination, as members of the project steering group, and general support:

Jules Agate, Jess Allen, John Baker, Helly Barber, Alys Black, Stephanie Boocock, Cliff Bradley, Valerie Bradley, Andrew Brett, Nick Button, James Byrne, George Cebo, Angela Charlton, Rosemary Chrimes, Rebecca Collins, Steve Coney, Nicky Davies, David Delaney, Jeremy Evans, Tom Fairfield, Fern Fellowes, Francis Flannigan, Garth Foster, Dave Gill, Felicity Gill, D. Griffin, Francesca Griffith, Trevor Griffiths, Alan Grigg, Leo Gubert, Martin Hales, Kate Hand, Nigel Hand, Sara Harris, Margaret Hawkins, Fergus Henderson, Brian Hicks, Hilary Hillier, Trevor Hulme, Hayley Jack, Patricia Jackson, Clive Jermy, Valerie Jermy, Malcolm John, Anna Jones, Rose Kibble, Phyl King, Richard King, Celia Kirby, Michael Lambert, Roger Lowery, Fiona Mac, Judy Malet, Neil Millington, Andrew Nixon, Louise Parker, John Partridge, Steve Redfern, Lydia Robbins, Cherilyn Roberts, Pat Robshaw, Steve Roe, Mary Scott, Darran Sharp, Rachel Sharp, David Smith, Hilary Smith, Colin Summers, Kim Summers, Bridgit Symons, Daniel Taylor, Jane Thomas, Bill Thompson, Will Watson, Stuart Webb, Valerie Webb, Dan Wenczek, Stephen West, Arthur Wild, Bronwen Wild, Ann Wilkinson, Jim Wilkinson, Pam Williams, Steve Wood, Ian Wrenn, Margaret Wrenn, Dorothy Wright (HCT).

The following have provided records in the past which are included in the species distribution maps:

Herefordshire Nature Trust's Community Biodiversity, Community Commons and other projects, Environmental Consultants, Hereford Museum, The National Trust, Herefordshire FWAG, English Nature, Herefordshire Council Ecologists, Anthea Brian and Beryl Harding and the Woolhope Naturalists' Club, Don Goddard, Gerald Leighton, Martin Noble, Ben Proctor, Jane Sweetman, the Herpetological Conservation Trust and the NBN.

HART is very grateful to all those Herefordshire Biological Records Centre volunteers who have given freely of their time to computerise the records for this book:

Doreen Beck, Roger Beck, Phyl King, Richard King, Steve Watkins, Heather Webster, Arthur Wild, Ann Wilkinson, Paul Zagni.

Advice and guidance was generously provided by:

Dr Anthea Brian MBE, Herefordshire Nature Trust; Prof. Garth Foster, Secretary of the Balfour-Browne Club; Janet Lomas and Mike Williams, Herefordshire Farming and Wildlife Advisory Group; Moira Jenkins, Herefordshire and Worcestershire Earth Heritage Trust; Tony King, designer of the pond database; Charlotte Morgan, Rural Development Agency; Dr. Keith Ray, County Archaeologist for Herefordshire; Rebecca Roseff, archivist and archaeologist; and Philip Warren, Sheffield University.

Amphibians and Reptiles and the law

The law protects all reptiles and amphibians to some extent.

Viviparous lizard, slow-worm, adder and grass snake are protected against intentional killing, injury and unlicensed trade in the animals (or any part thereof) under the Wildlife and Countryside Act 1981 (as amended). It is also illegal to keep adders in captivity without the necessary licence as required by the Dangerous Wild Animals Act 1976.

Common frog, common toad, smooth and palmate newt (or any part thereof) are all protected against sale, trade, barter or exchange, under the Wildlife and Countryside Act 1981.

The great crested newt and its habitat are fully protected under the Wildlife and Countryside Act 1981 (as amended), and the Habitats Regulations 1994. It is illegal to:

- Intentionally or deliberately capture, kill or injure great crested newts.
- Intentionally, deliberately or recklessly* damage, destroy or obstruct access to any place used for shelter or protection, including resting or breeding places (occupied or not).
- Intentionally, deliberately or recklessly* disturb great crested newts when in place of shelter.
- Sell, barter, exchange or transport, or offer for sale great crested newts or parts thereof.

This legislation covers all life stages (eggs, larvae, juveniles and adults).

* Recklessly applies to England and Wales as an amendment to the Wildlife and Countryside Act 1981 as a result of the CRoW Act 2000.

A licence is required from Natural England or the appropriate statutory nature conservation agency to carry out surveys for scientific or educational purposes involving disturbance and handling of the great crested newt.

Habitat and the law

Site of Special Scientific Interest (SSSI) status confers some degree of protection from potentially dangerous operations. Guidelines suggest that reptile SSSIs should be selected by the following criteria:

- In any 'area of search', usually defined by county or vice-county boundary, the best locality containing at least three of the common species, adder, grass snake, viviparous lizard and slow-worm should be selected.

- Sites should not be chosen just to represent populations of one or two species, but the occurrence of any species should count positively in the evaluation of sites chosen largely on other grounds, especially in areas where the species concerned is rare or at the geographical limit of its range.

All exceptional great crested newt sites are eligible for SSSI status. Exceptional sites are considered to be those where a night-time torch count in the breeding season exceeds a hundred individuals. The other amphibian species are considered to be widespread and relatively numerous, so the presence of an outstanding assemblage of these is the guideline for site selection.

Biodiversity Action Plans (BAPs) in the UK

Following the 1992 Earth Summit in Rio, at which one of the key agreements was the Convention on Biological Diversity, the UK published its Biodiversity Action Plan in 1994. It currently contains 391 species action plans and 45 habitat action plans. These plans however are dependent on action at the local level.

While not providing any legal protection, UK and local BAPs aim to promote focussed conservation action for the most endangered species and habitats.

The Herefordshire Biodiversity Partnership published the Herefordshire Local BAP (LBAP) in 2000, with an update in 2005. 76% of the targets set in the LBAP for conservation of key species in Herefordshire have been met in the first five years. Herefordshire Council's Community Strategy recognises the importance of the LBAP for biodiversity and the environment. In the LBAP each species plan has a lead partner e.g. a wildlife organisation or local voluntary group, and HART is the lead for the adder and the great crested newt.

For more information, contact:

The Herefordshire Biodiversity Officer, Herefordshire Biodiversity Partnership, 01432 383026. www.herefordbap.org.uk

Development on amphibian and reptile sites

Reptiles and amphibians may be threatened and the law potentially breached by development activities. To keep within the law developers and their ecological consultants should check the current legislation. Such information is widely available, including from the Planning Department of Herefordshire Council, Natural England, Froglife, the Herpetological Conservation Trust, and HART. The County Amphibian and Reptile Recorders or the HBRC can provide details of reptile and amphibian occurrence.

The laws can and do change so it is advisable to keep up to date with changes. The Quinquennial Review, a statutory 5-year review procedure carried out by the Joint Nature Conservation Committee (JNCC) reviews eligibility of native species in the UK to be added to or removed from schedules 5 and 8 of the Wildlife and Countryside Act. The first Quinquennial review in 1988 added the grass snake, slow-worm and viviparous lizard from killing and injuring, the adder was added to schedule 5 in 1991.